T0325909

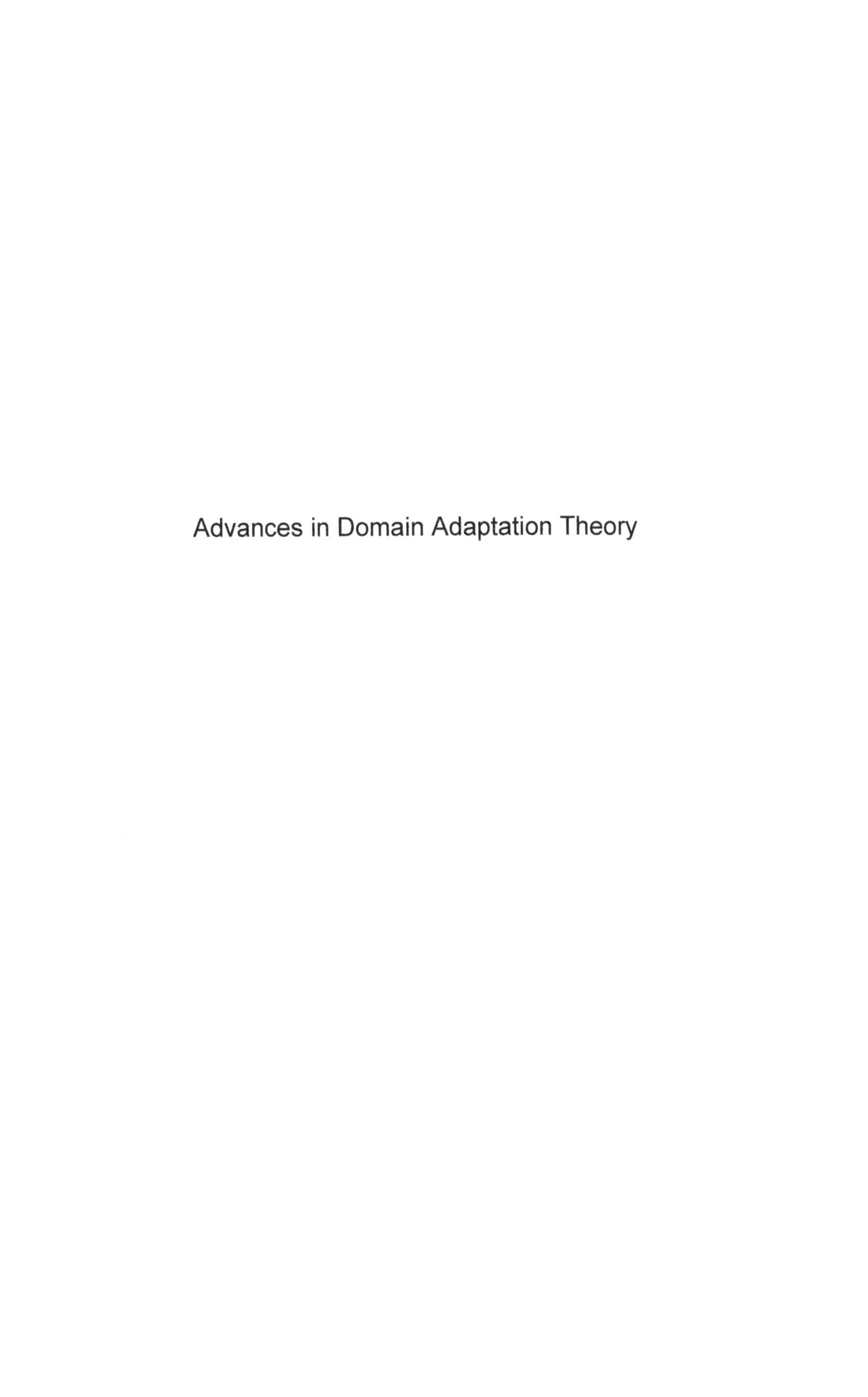

Advances in Domain Adaptation Theory

Series Editor
Jean-Charles Pomerol

Advances in Domain Adaptation Theory

Ievgen Redko
Amaury Habrard
Emilie Morvant
Marc Sebban
Younès Bennani

First published 2019 in Great Britain and the United States by ISTE Press Ltd and Elsevier Ltd

ISTE Press Ltd
27-37 St George's Road
London SW19 4EU
UK

www.iste.co.uk

Elsevier Ltd
The Boulevard, Langford Lane
Kidlington, Oxford, OX5 1GB
UK

www.elsevier.com

British Library Cataloguing-in-Publication Data
A CIP record for this book is available from the British Library
Library of Congress Cataloging in Publication Data
A catalog record for this book is available from the Library of Congress
ISBN 978-1-78548-236-6

Printed and bound in the UK and US

Contents

Abstract

All famous machine learning algorithms that correspond to both supervised and semi-supervised learning work well only under a common assumption: training and test data follow the same distribution. When the distribution changes, most statistical models must be reconstructed from new collected data that may be costly or even impossible to get for some applications. Therefore, it became necessary to develop approaches that reduce the need and the effort demanded for obtaining new labeled samples, by exploiting data available in related areas and using it further in similar fields. This has given rise to a new family of machine learning algorithms, called transfer learning: a learning setting inspired by the capability of a human being to extrapolate knowledge across tasks to learn more efficiently. Despite a large number of different transfer learning settings, the main objective of this book is to provide an overview of the state-of-the-art theoretical results in a specific and arguably the most popular subfield of transfer learning, called domain adaptation.

In order to highlight the difference between traditional statistical learning theory and domain adaptation theory, we first provide a brief overview of the most popular concepts used to derive traditional generalization guarantees in Chapter 1. This overview includes Vapnik–Chervonenkis, Rademacher, probably approximately correct (PAC)–Bayesian, robustness and stability-based bounds. Then, in Chapter 2, we introduce and explain the problem of domain adaptation. This introduction to domain adaptation is followed by the description of four major families of theoretical results existing in the literature. The first family considered is the *divergence-based* generalization bounds including the results based on $\mathcal{H}\Delta\mathcal{H}$-divergence, discrepancy distance (Chapter 3) and integral probability metrics (Chapter 5). The second family of considered learning bounds are the *PAC Bayesian bounds* (Chapter 6). These bounds include the original PAC–Bayesian results for domain adaptation and their recent updated version which provides a new perspective on the former. The latter two chapters are further followed by a presentation of the generalization guarantees based on the robustness and stability properties of learning algorithms (Chapter 7).

Finally, we present the impossibility and hardness theorems for domain adaptation (Chapter 4), establishing the pitfalls of domain adaptation algorithms, as well some necessary and sufficient conditions required for them to succeed (Chapter 8). We conclude this book by summarizing the main existing results in the area of domain adaptation and by providing some hints for important future investigations and open problems.

Notations

$\mathbf{X}$	Input space
Y	Output space
$\mathcal{D}$	A domain: a yet unknown distribution over $\mathbf{X} \times Y$
$\mathcal{D}_{\mathbf{X}}$	Marginal distribution of $\mathcal{D}$ on $\mathbf{X}$
$\mathcal{S}$	Source domain (i.e. source distribution over $\mathbf{X} \times Y$)
$\mathcal{T}$	Target domain (i.e. target distribution over $\mathbf{X} \times Y$)
$\mathcal{S}_{\mathbf{X}}$	Marginal distribution of $\mathcal{S}$ on $\mathbf{X}$
$\mathcal{T}_{\mathbf{X}}$	Marginal distribution of $\mathcal{T}$ on $\mathbf{X}$
$\hat{\mathcal{S}}_{\mathbf{X}}$	Empirical distribution associated with a sample drawn from $\mathcal{S}_{\mathbf{X}}$
$\hat{\mathcal{T}}_{\mathbf{X}}$	Empirical distribution associated with a sample drawn from $\mathcal{T}_{\mathbf{X}}$
$\textsc{supp}(\mathcal{D})$	Support of distribution $\mathcal{D}$
$\mathbf{x} = (x_1, \ldots, x_d)^\top \in \mathbb{R}^d$	A d-dimensional real-valued vector
$(\mathbf{x}, y) \sim \mathcal{D}$	$(\mathbf{x}, y)$ is drawn i.i.d. from $\mathcal{D}$
$S = \{(\mathbf{x}_i, y_i)\}_{i=1}^m \sim (\mathcal{D})^m$	Labeled learning sample consisting of m examples drawn i.i.d. from $\mathcal{D}$
$S_u = \{(\mathbf{x}_i)\}_{i=1}^m \sim (\mathcal{D}_{\mathbf{X}})^m$	Unlabeled learning sample consisting of m examples drawn i.i.d. from $\mathcal{D}_{\mathbf{X}}$
$\|S\|$	Size of the set S
$\mathcal{H}$	Hypothesis space
π	Prior distribution on $\mathcal{H}$
ρ	Posterior distribution on $\mathcal{H}$
$\mathrm{KL}(\cdot\|\cdot)$	Kullback–Leibler divergence
$\mathbf{I}[a]$	Indicator function: returns 1 if a is true, 0 otherwise
$\mathrm{sign}[a]$	Return the sign of a: 1 if $a \geq 0$, -1 otherwise
$\mathbf{M}$	An arbitrary matrix
$\mathbf{M}^\top$	Transpose of the matrix $\mathbf{M}$
$\mathbf{0}$	Null matrix

$\langle \cdot, \cdot \rangle$	Dot product between two vectors
$\|\cdot\|_1$	L_1-norm
$\|\cdot\|_2$	L_2-norm
$\|\cdot\|_\infty$	L_∞-norm
$\mathbf{Pr}\left(\cdot\right)$	Probability of an event
$\mathbf{E}\left(\cdot\right)$	Expectation of a random variable
$\mathbf{Var}\left(\cdot\right)$	Variance of a random variable

Introduction

Machine learning[1] is a computer science discipline generally considered as one of the fields of study of artificial intelligence that is located at the frontier of computer science and applied mathematics (statistics and optimization). The notion of learning encompasses any algorithmic procedure that builds a mathematical model from observed data for making decisions on new unseen data. The learning objective of such an algorithm is thus to automatically extract and exploit the information contained in the data, and to focus on the development, analysis and implementation of methods aimed at extracting useful knowledge from observations given as input. Therefore, learning algorithms may solve very different problems and come in a large variety of forms. Among others, we can mention a popular classification problem, one of the most famous illustrations of which is the automatic differentiation between spam and ham (i.e. the desired e-mails): given a collection of messages for a given user, the goal of the algorithm is to be able to extract the patterns from them, in order to classify new messages correctly. Another very popular example is the regression problem: a setting where we want to predict a certain value based on the collected observations, without it being restricted to a set of categories defined beforehand. For example, one may want to predict the temperature of the day from the information recorded by weather stations over the previous years. Depending on the data and objectives, many applications of machine learning exist including multimedia, biology, signal processing and automatic language processing.

In this book, we support the framework of statistical learning related to the theory of statistical studies on empirical processes introduced by Vladimir Vapnik and Alexey Chervonenkis. In particular, we are interested in the paradigm of *supervised* learning where the observed data are labeled. This is contrary to the unsupervised learning for which no such supervision of observed data is available. Let us now take the classic

1 The reader can refer to [MIT 97, BIS 06, MOH 12, SHA 14] for books on general machine learning.

and concrete examples of a spam/ham filtering system. In this example, we have at our disposal observations given by already labeled messages in the form of couples $(\mathbf{x}, y)$, where $\mathbf{x}$ is the representation of a message by a vector of word frequencies in the text and y is the class/label assigned to the message $\mathbf{x}$. In our case, the class/label can take only two values that are given by -1 or $+1$, so that -1"=" spam, $+1$"=" ham. The objective is then to find (to learn) a function that automatically assigns one of these two classes to a new message and that commits as few errors as possible over time. This is called classification because the task is clearly to predict the class to which a data instance belongs, and it is supervised because the learning is based on already labeled observations.

In the literature, many algorithms have been proposed to solve this type of classification task. The most intuitive is probably the k nearest neighbor method (k-NN) [COV 67], which simply consists of choosing the majority class from the k observations closest to the data point to be labeled. Another famous algorithm is support vector machines (SVM) [BOS 92], which aim to learn a classifier by maximizing the distance between the line separating the classes of interest and the observations from each of these classes. This latter concept, commonly called the *margin*, makes it possible to define a notion of confidence of the data classification: one is confident about the predictions made for points having a large margin with respect to the obtained linear classifier. While learning a linear classifier is only efficient when classes are linearly separable, the SVM approach can also be extended to learn nonlinear classifiers by reformulating the problem in the form of a linear classification in the projection space defined, using a particular kind of nonlinear functions called a kernel. Another popular type of method is the neural network approach[2], inspired by biological neural networks. A neural network generally consists of a succession of interconnected layers of so-called neurons, whose inputs correspond to the outputs of the previous layer: each neuron is connected to the neurons of the previous layer by synapses classically calculating a weighted sum of the outputs of the previous layer. One of the simplest historical algorithms that falls into this category is that of the perceptron [ROS 58]. Recent extensions of these approaches form an active and very important machine learning topic called *deep learning* [BEN 09]. As mentioned above, different learning methods have already found their application in a wide variety of areas of our everyday life, sometimes even without us being aware of it. For instance, machine learning methods for speech recognition are often integrated into our smartphones in order to enable the understanding of human voice, allowing us to interact with it. The learning algorithms that form an active field of recommendation systems are often used to provide us with suggestions on what we would like to choose, based on our preferences on different websites such as Amazon, Booking or IMDB. Various pattern recognition methods are used in an extremely vast range of applications, varying from tumor detection in a human brain to spotting wildlife in remote parts of

2 The foundations of neural networks approaches have been introduced by [LET 59].

Africa. All these examples are part of a constant trend in which statistical learning is taking an increasingly important place in our lives and is undoubtedly becoming irreplaceable.

From the methodological point of view, supervised classification is usually formalized as the learning or construction of a function (often called hypothesis or classifier, from a learning sample composed of independent observations) and is identically distributed according to a fixed and unknown probability distribution. Such a model can take the form of a hyperplane capable of correctly separating new objects. An important aspect must be underlined here: the learned model must be efficient on observations but also – and especially – on new unseen data. In other words, since the data distribution is assumed to be unknown, the major issue in statistical learning is the learning of a classifier that must have good *generalization properties*. The intuition behind this process is illustrated in Figure I.1.

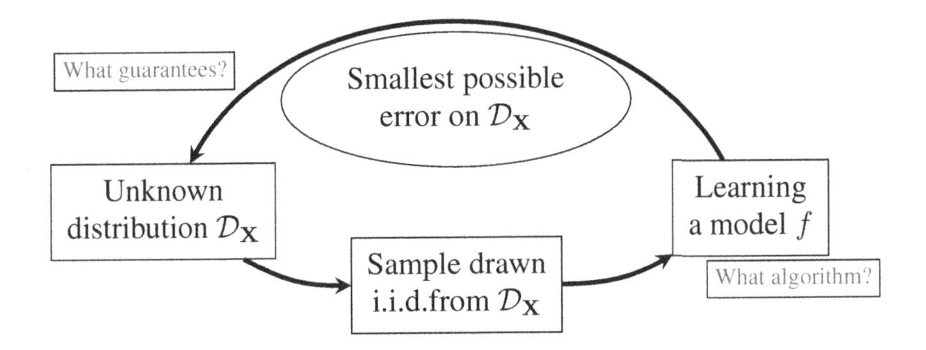

Figure I.1. *Intuition of problems related to automatic learning: given a task, by what method and with what guarantees can we learn a successful model?*

The notion of generalization plays a crucial role in machine learning, as it allows us to understand to what extent the knowledge extracted from observations by a learning algorithm from the available observations characterizes the statistical properties of the unknown distribution from which they were drawn from. For example, consider the classification task where our goal is to differentiate images of cats from those of dogs. A possible naive strategy to solve this problem is to explicitly memorize all the cat images (if the sample size allows it). This approach will almost certainly lead to a perfect classification of the observed images, because in this case the object classes are memorized by heart. However, this naive learning algorithm would fail on any other sample drawn from the same distribution, unless it encounters an image that it has already memorized before. Another more intelligent learning strategy would be to extract from the available images a set of characteristics that differentiate cats and dogs according to their physiological traits. Unlike the above-mentioned naive approach, this strategy may be potentially less efficient on the original sample of images, but it will probably be able to process any new image sample because the physiological differences between the two classes will remain the same, regardless of the example given as input. This simple illustration gives us a hint not only about the generalization property of a given learning algorithm and the importance of this former, but also about the general shape of theoretical guarantees that one seeks to obtain. The foremost goal of any theoretical guarantee is to show that learning from a finite-size sample S drawn from a data distribution $\mathcal{D}_{\mathbf{X}}$ ensures, with a high probability, effective learning on any possible sample from the same distribution. This kind of result is expressed as an inequality for which our first take may look something like this:

$$\begin{array}{c}\text{classification error over}\\ \text{any sample from } \mathcal{D}_{\mathbf{X}}\end{array} \leq \begin{array}{c}\text{classification error}\\ \text{for a given sample } S\end{array}.$$

However, from the example mentioned above, we know that a learning strategy that memorizes labels of the samples is expected to fail in the long run, as its decision rules regarding the assignment of a label to a new image will require comparing it to all previously seen images. This is contrary to our second approach that extracts meaningful physiological characteristics capturing a general knowledge regarding cats and dogs, where a comparison only with respect to several characteristics will be made. To this end, a meaningful generalization result will require the incorporation of an overall so-called "complexity" of the model produced by a given learning algorithm. Thus, we can write our second take at the generalization guarantees as follows:

$$\begin{array}{c}\text{classification error over}\\ \text{any sample from } \mathcal{D}_{\mathbf{X}}\end{array} \leq \begin{array}{c}\text{classification error}\\ \text{for a given sample } S\end{array} + \begin{array}{c}\text{complexity of the}\\ \text{learning algorithm}\end{array}.$$

Here, the term related to the complexity of the model produced by a given learning algorithm can be seen as the complexity of the rules used by the model to make its decision regarding the label of each instance. On the one hand, a model that memorizes images and labels by heart requires a pixel-wise comparison of the given image with each image it has memorized. This process, obviously, is extremely complex and prohibitive. On the other hand, the classification of cats and dogs according to their physiological characteristics only requires the comparison of the discriminant parts of the images that highlight the difference between the two classes of interest. In this case, the complexity of the model is much lower. However, it should be noted that in both cases, the size of the learning sample S also has to play a very important role. To illustrate this, let us suppose that our learning sample consists of all the cats and dogs on planet Earth and that our naive model will never encounter an image of a dog or cat that it has not already memorized. The two models should then have a very high generalization capacity because the naive model will be able to memorize everything it will have to classify in the future, and the second will have access to all the information on the differences between the two species to avoid any confusion. This observation leads to a slight modification of the last inequality, which now takes the following form:

$$\begin{matrix}\text{classification error over} \\ \text{any sample from } \mathcal{D}_{\mathbf{X}}\end{matrix} \leq \begin{matrix}\text{classification error} \\ \text{for a given sample } S\end{matrix} + \frac{\begin{matrix}\text{complexity of the} \\ \text{learning algorithm}\end{matrix}}{\begin{matrix}\text{the size of the} \\ \text{learning sample } S\end{matrix}}.$$

This last inequality presents a general result of the learning theory capturing the intuition behind the conditions that have to be fulfilled in order to lead to a high generalization capacity. In practice, the learning algorithm must minimize the classification error on the learning sample S while ensuring that its decision-making model is not too complicated, even at the cost of a few additional errors on S. Finally, increasing the size of S allows us to bound in the tightest way possible the error on any sample coming from the distribution $\mathcal{D}_{\mathbf{X}}$. As we show in the Chapter 1, this kind of result can be proven in a variety of ways. Some of the theoretical results discussed in this book are related to the theoretical complexity of the model, whereas others are data dependent and can be calculated explicitly according to the available instances. Other generalization results are based on certain properties of the algorithms related to their "robustness" or "stability". All these contributions can be roughly united under the name "statistical learning theory".

Unsurprisingly, the generalization properties as a field of study attracts a lot of attention from scientists in the area of machine learning. This interest is confirmed by two widely recognized international conferences, the Conference on Learning Theory (COLT) and the Conference on Algorithmic Learning Theory (ALT), where the majority of results presented refer, in one form or another, to the generalization capacities of a learning model. This trend is also widespread in all other major conferences on the machine learning field where the subject "learning theory" inevitably appears among the list of covered topics and the vast majority of presented contributions come together with their generalization guarantees.

1

State of the Art of Statistical Learning Theory

As we have previously seen, the general purpose of machine learning is to automate the process of acquiring knowledge based on observed data. Despite this general objective, the way in which this process is carried out may vary considerably depending on the nature of considered data and the final goal of learning. Thus, it can be characterized in many completely different ways. What remains unchanged, however, is that the nature of the learning paradigms. Despite their apparent variety, they can be roughly called "statistical", which means that they are all related to revealing the statistical nature of the underlying phenomenon. To this end, the main goal of *statistical learning theory* is to analyze and formalize the process of knowledge acquisition and to provide a theoretical basis and an appropriate context for its analysis within a statistical framework.

Statistical learning theory has a rich history and continues to constantly broaden the horizons of machine learning practitioners and responds to their needs by analyzing new and emerging machine learning methods. The motivation to pursue theoretical research in the field of machine learning can be justified by Vladimir Vapnik's famous expression:

"NOTHING IS MORE PRACTICAL THAN A GOOD THEORY".

As our book is related to a particular topic in statistical learning theory, it would have been incomplete without an introduction to its basics and a brief description of its state-of-the-art results.

This chapter describes the classic statistical machine learning context in which the domain adaptation theory presented in this book stands. As mentioned in the Introduction, we focus here on the classic framework of supervised classification, in which one usually aims to learn a function capable of expressing the relationship between an input space (the representation of the data) and an output space (a category or class). Given a task, one considers a sample of observed data (the training sample) and a set of possible functions (the hypothesis space). Then, the main goal is to find a function that is consistent with the training data and that is general enough to have. Statistical learning theory, which is a popular approach to providing such a type of analysis was introduced by Vladimir Vapnik in the 1970s and by Leslie Valiant in the 1980s.

More precisely, we formalize the usual supervised classification setting and introduce the notations used throughout our book in section 1.1. In section 1.2, we recall the seminal works on probably approximately correct (PAC) bounds [VAL 84], Vapnik–Chervonenkis (VC) bounds [VAP 06, VAP 71], Rademacher complexity [KOL 99], and PAC–Bayesian theory [MCA 99]. These families of results laid the foundation of statistical learning theory. We then continue our presentation with more recent results proposed based on the notion of algorithmic stability [BOU 02] and algorithmic robustness [XU 10].

1.1. Preliminaries

In this section, we introduce the definitions used to present major types of theoretical results that can be found in the statistical learning literature.

1.1.1. *Definitions*

Let a pair $(\mathbf{X}, Y)$ define the input and the output spaces that will also be, respectively, called a feature space and a labeling set. We consider the case where all data given by $\mathbf{X}$ are described by real-valued vectors of finite dimension d, i.e. $\mathbf{X} \subseteq \mathbb{R}^d$. We note that depending on the application, both $\mathbf{X}$ and Y can be defined in different ways: while $\mathbf{X}$ may not exclusively contain vector-valued data but also structural or relational data; the output space Y, in its turn, may be continuous, discrete or structured. Regarding $\mathbf{X}$, we limit ourselves only to the case described above; as for Y, we distinguish between two possible scenarios:

– when Y is continuous, e.g. $Y = [-1, 1]$ or $Y = \mathbb{R}$, we talk about regression;

– when Y is discrete and takes values from a finite set, we talk about classification. Two important cases of classification are binary classification and multiclass classification where $Y = \{-1, 1\}$ (or $Y = \{0, 1\}$) and $Y = \{1, 2, \ldots, C\}$ with $C > 2$, respectively.

Now, given $\mathbf{X}$ and Y and assuming that there is an unknown joint probability distribution $\mathcal{D}$ defined on their product space $\mathbf{X} \times Y$, we introduce the following quantities:

1) $S = \{(\mathbf{x}_i, y_i)\}_{i=1}^m \sim (\mathcal{D})^m$ is a training sample (also called a *learning sample*) of m independent and identically distributed (i.i.d.) pairs (also called examples or data instances) $(\mathbf{x}_i, y_i) \in (\mathbf{X}, Y)$ according to $\mathcal{D}$.

2) $\mathcal{H} = \{h | h : \mathbf{X} \to Y\}$ is a *hypothesis space* (also called *hypothesis class*) consisting of functions that map each element of $\mathbf{X}$ to Y. These functions h are usually called hypotheses, or more specifically classifiers or regressors depending on the nature of Y.

Given the training sample $S \sim (\mathcal{D})^m$, the goal of a learner usually consists of finding a good hypothesis function $h \in \mathcal{H}$ that captures in the best way possible the relationship between $\mathbf{X}$ and Y. This relationship, however, often extends beyond the observed sample S to unseen pairs $(\mathbf{x}, y)$ drawn from the same probability distribution $\mathcal{D}$. In what follows, we assume that the future (i.e. test) observations are related to the past (i.e. training) ones and that training and test instances are drawn independently from the same distribution (i.i.d.). Here, the independence nature of sampling means that each new observation should provide maximum information, while the identical distribution means that the observations are informative with respect to their underlying probability distribution. In order to measure how well the hypothesis reveals this relationship, we consider a loss function $\ell : Y \times Y \to [0, 1]$ that gives a cost of $h(\mathbf{x})$ deviating from the true output $y \in Y$. Then, to capture the notion of the loss over all samples coming from some probability distribution $\mathcal{D}$, we define the *true risk* below.

DEFINITION 1.1 (True risk).– *Given a loss function $\ell : Y \times Y :\to [0, 1]$, the true risk (also called generalization error) $\mathrm{R}_{\mathcal{D}}^{\ell}(h)$ for a given hypothesis $h \in \mathcal{H}$ on a distribution $\mathcal{D}$ over $\mathbf{X} \times Y$ is defined as*

$$\mathrm{R}_{\mathcal{D}}^{\ell}(h) = \mathop{\mathbf{E}}_{(\mathbf{x},y)\sim\mathcal{D}} \ell(h(\mathbf{x}), y).$$

By abusing the notation, for a given pair of hypotheses $(h, h') \in \mathcal{H}^2$, we write

$$\mathrm{R}_{\mathcal{D}}^{\ell}(h, h') = \mathop{\mathbf{E}}_{(\mathbf{x},y)\sim\mathcal{D}} \ell(h(\mathbf{x}), h'(\mathbf{x})).$$

This definition introduces a new important property that is called the generalization capability of a hypothesis. As we can see, the true risk is defined as the expected value over the whole product space $\mathbf{X} \times Y$ and thus it measures the expected error over any other previously unseen samples from $\mathcal{D}$ as well. Naturally, the best hypothesis is the one that minimizes this true risk and that, consequently, generalizes well.

However, in practice we do not know the true distribution $\mathcal{D}$, making the calculation of the true risk for a given hypothesis impossible. Therefore, it is natural to define the notion of empirical risk, which reflects the error of some hypothesis h obtained on a fixed observable training sample S.

DEFINITION 1.2 (Empirical risk).– *Given some loss function* $\ell : Y \times Y :\to [0,1]$ *and a training sample* $S = \{(\mathbf{x}_i, y_i)\}_{i=1}^m$ *where each example is drawn i.i.d. from* $\mathcal{D}$*, the empirical risk* $\mathrm{R}^{\ell}_{\hat{\mathcal{D}}}(h)$ *for a given hypothesis* $h \in \mathcal{H}$ *is defined as*

$$\mathrm{R}^{\ell}_{\hat{\mathcal{D}}}(h) = \frac{1}{m} \sum_{i=1}^{m} \ell(h(\mathbf{x}_i), y_i),$$

where $\hat{\mathcal{D}}$ *is an empirical distribution associated with sample* S.

In what follows, we use the notations $\mathrm{R}^{\ell}_{\hat{\mathcal{D}}}(h)$ and $\mathrm{R}^{\ell}_{S}(h)$ interchangeably where the former generally refers to an empirical error on some arbitrary finite sample drawn from $\mathcal{D}$, while the latter is deployed when the sample S is defined explicitly. In these definitions, we make reference to the notion of a loss function ℓ given by a function that stands for a dissimilarity measure between two labels. The most natural loss function that can be used to count the number of errors committed by a hypothesis function $h \in \mathcal{H}$ on the distribution $\mathcal{D}$ is $0-1$ loss function $\ell_{0-1} : Y \times Y \to \{0,1\}$, which is defined for a training example $(\mathbf{x}, y)$ as

$$\begin{aligned} \ell_{01}(h(\mathbf{x}), y) &= \mathbf{I}\left[h(\mathbf{x}) \neq y\right] \\ &= \begin{cases} 1, & \text{if } h(\mathbf{x}) = y, \\ 0, & \text{otherwise.} \end{cases} \end{aligned} \qquad [1.1]$$

However, it can be proved that minimizing $0-1$ loss is an NP-hard task and thus other approximations of this intuitive loss function should be considered (one often talks about surrogate loss). One of them is the hinge loss, known as the best approximation of the $0-1$ loss [BEN 12a], defined for a given pair $(\mathbf{x}, y)$ by

$$\begin{aligned} \ell_{\text{hinge}}(h(\mathbf{x}), y) &= [1 - yh(\mathbf{x})]_+ \\ &= \max(0, 1 - yh(\mathbf{x})). \end{aligned}$$

Another loss function often used in practice that extends $0-1$ loss to the case of real values is linear loss $\ell_{\text{lin}} : \mathbb{R} \times \mathbb{R} \to [0,1]$, defined as follows:

$$\ell_{\text{lin}}(h(\mathbf{x}), y) = \frac{1}{2}(1 - yh(\mathbf{x})).$$

The three above-mentioned loss functions are illustrated in Figure 1.1. As we will see in the following sections, one of the key ingredients in finding a good hypothesis function is the minimization of the empirical error.

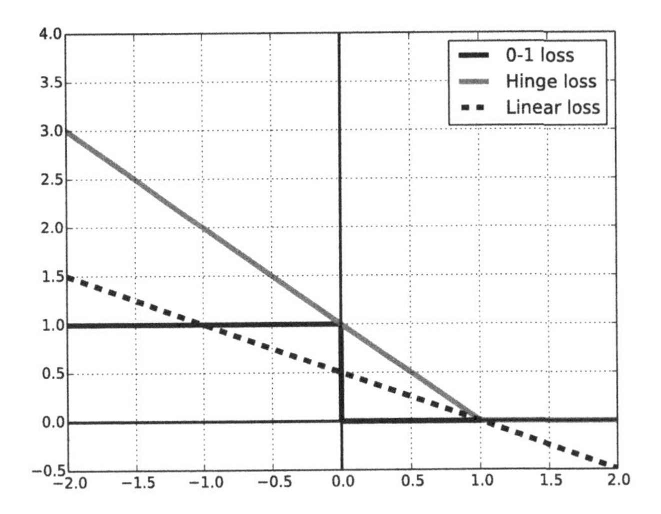

Figure 1.1. *Illustration of the $0-1$ loss (in black), the linear loss (dashed line) and the hinge loss (in gray)*

1.1.2. *No-free lunch theorem*

One may note that the i.i.d. assumption suggests that every data instance carries some new information that is independent from other previously seen samples. This may suggest the existence of a consistent universal learner returning a hypothesis that becomes closer and closer to the perfect one when the sample size increases. This, however, is not true as any consistent learner can have an arbitrary bad behavior on a sample of finite size. This statement is usually formalized by the means of the famous "no-free lunch" theorem that can be stated as follows.

THEOREM 1.1.– *Let $\mathcal{A}$ be any learning algorithm for the task of binary classification with respect to the $0-1$ loss over a space $\mathbf{X}$. Let m be any number smaller than $|\mathbf{X}|/2$, representing a training set size. Then, there exists a distribution $\mathcal{D}$ over $\mathbf{X} \to \{0,1\}$ such that:*

1) there exists a function $h : \mathbf{X} \to \{0,1\}$ with $\mathrm{R}^{\ell}_{\mathcal{D}}(h) = 0$;

2) with probability of at least $\frac{1}{7}$ over the random choice of $S \sim (\mathcal{D})^m$, we have that $\mathrm{R}^{\ell}_{\mathcal{D}}(\mathcal{A}(S)) \geq \frac{1}{8}$.

As mentioned above, this theorem shows that there is no universal learner that succeeds on all tasks. Consequently, every learner has to be designed for a specific task by using prior knowledge about that task in order to succeed. Below we present several strategies that can be used to obtain a hypothesis with good generalization capacities and low error based on choosing a "good" hypothesis class and on restricting this latter to have a reasonable complexity.

1.1.3. *Risk minimizing strategies*

Obviously, in the extreme case where one possesses a training sample of infinite size $m \to +\infty$, the best possible strategy to find a good hypothesis is to minimize the empirical risk, as it approximates correctly the true risk and thus generalizes well. In reality, when we deal with a training sample of a finite size, the strategy consisting of finding the best hypothesis by minimizing only the empirical risk has a significant drawback associated with *overfitting*. Indeed, learning a hypothesis that is tailor-made to perfectly make predictions on the observable samples corresponding to a "memorization by heart" of the training data rather than an "interpolation" of the concepts that will work to generalize well on unseen data coming from the same probability distribution. The problem of overfitting often arises when one considers a hypothesis class $\mathcal{H}$ constituted by overly complex functions that allow the model to be fitted to training instances (that do not sufficiently reflect the true population's distribution). This situation is depicted in Figure 1.2.

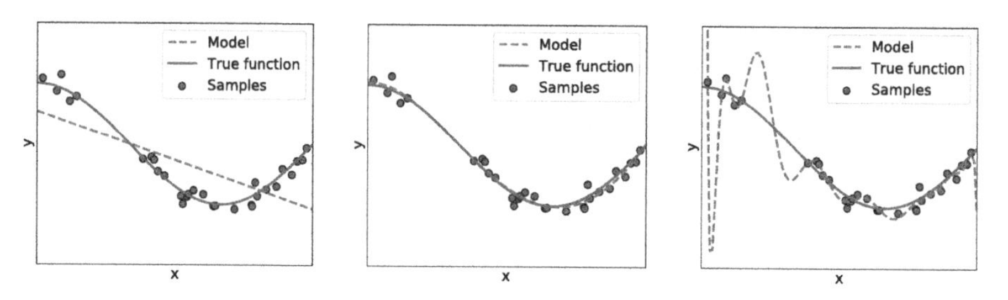

Figure 1.2. *Illustration of underfitting versus overfitting. (Left) the model is underfitted and does not properly capture the true behavior of the data sample; (middle) a good model that follows the true distribution function of the data sample; (right) the model is overfitted and tries to follow perfectly the points of the available sample. For a color version of this figure, see www.iste.co.uk/redko/domain.zip*

To overcome this issue, one should consider a risk minimization strategy that finds a good trade-off between the complexity of the hypothesis space and the minimization of the empirical risk. There are several well-known strategies that allows us to find a hypothesis function with low true risk, namely:

1) empirical risk minimization (ERM);

2) structural risk minimization (SRM);

3) regularized risk minimization (RRM).

We now consider each of them in the following.

1.1.3.1. *Empirical risk minimization*

We start with the most straightforward way to proceed, which is to directly minimize the empirical risk based on the training sample. More formally, given a training sample $S \sim (\mathcal{D})^m$, we pick up from the hypothesis class $\mathcal{H}$ an optimal hypothesis h_S^* such that

$$h_S^* = \underset{h \in \mathcal{H}}{\text{argmin}} \ \mathrm{R}_S^\ell(h),$$

and hope that $\mathcal{H}$ was constrained beforehand to a reasonable class of functions using some *a priori* knowledge about the learning problem. However, when this knowledge is not available, this strategy may be subject to overfitting due to the reasons explained before.

1.1.3.2. *Structural risk minimization*

As defining the hypothesis class beforehand is usually quite hard, one may rather consider using a sequence of hypothesis classes $\mathcal{H}_j$ with $j = 1, 2, \ldots$ ordered by their complexity so that $\forall\, j = \{1, 2, \cdots\}$, $\mathcal{H}_j \subseteq \mathcal{H}_{j+1}$. This idea of structural risk minimization, first set out in [VAP 71], further consists of finding a hypothesis function that minimizes the empirical risk over the whole sequence of hypothesis classes considered. This leads to the following procedure:

$$h_S^* = \underset{h \in \mathcal{H}_j j = \{1,2,\ldots\}}{\text{argmin}} \left\{ \mathrm{R}_S^\ell(h) + \text{pen}(\mathcal{H}_j) \right\},$$

where the term $\text{pen}(\mathcal{H}_j)$ penalizes hypothesis class $\mathcal{H}_j$ based on its complexity. This approach gives us a possibility to define in a formal way how the trade-off between the complexity of a hypothesis class and the performance of a given hypothesis function from it should be related. Unfortunately, it raises some other issues related to the definition of the complexity of hypothesis classes and its computability for finite samples. We will talk about this later in this chapter.

1.1.3.3. *Regularized risk minimization*

Similarly to SRM, one may introduce a trade-off between finding a hypothesis with a low empirical risk and high generalization capacities by penalizing the hypothesis function itself. In this case, we can consider the penalty term as a sort of norm minimization that restricts the hypothesis' complexity. More formally, given a training sample $S \sim (\mathcal{D})^m$, we look for the best hypothesis h_S^* by $\mathcal{H}$ minimizing

$$h_S^* = \underset{h \in \mathcal{H}}{\text{argmin}} \ \left\{ \mathrm{R}_S^\ell(h) + \lambda \left\| h \right\| \right\},$$

where λ controls the trade-off between the two terms. One may note that standardizing the complexity of the hypothesis function can be achieved by codifying the main terms that define it. For instance, in regression this may mean regularizing the norm of the vector of coefficients in order to make sure that the resulting function will be defined only with respect to a restricted number of covariates. In general, we note that this approach is by far the most popular one, at least when it comes to the algorithmic implementations.

1.2. Generalization bounds

Based on the ideas introduced above, an important question that can be asked is to what extent the proposed minimization strategies can be justified from a statistical point of view. To this end, statistical learning theory [VAP 95] provides us with results regarding the conditions that lead to consistent estimators of the empirical risk that ensure their convergence to the value of the true risk. These results, known as *generalization bounds*, are usually expressed in the form of PAC bounds [VAL 84] that have the following form:

$$\mathop{\mathbf{Pr}}_{S\sim(\mathcal{D})^m}\left\{|\mathrm{R}^{\ell}_{S}(h) - \mathrm{R}^{\ell}_{\mathcal{D}}(h)| \leq \varepsilon\right\} \geq 1-\delta,$$

where $\varepsilon > 0$ and $\delta \in (0,1]$. The last expression essentially tells us that we want to upper bound the gap between the true risk and its estimated value by the smallest value of ε possible, this with a high probability over the random choice of the training sample S. The major question now is to understand whether $\mathrm{R}^{\ell}_{S}(h)$ converges to $\mathrm{R}^{\ell}_{\mathcal{D}}(h)$ with an increasing number of samples and what the speed of this convergence is. We now proceed to a presentation of several theoretical paradigms that were proposed in the literature in order to show different characteristics of a learning model or a data sample that this speed can depend on.

1.2.1. *Vapnik–Chervonenkis bounds*

VC bounds [VAP 71, VAP 06] are based on the original definition that allows us to quantify the complexity of a given hypothesis class. This notion of complexity is captured by the famous VC dimension defined as follows.

DEFINITION 1.3 (VC dimension).– *The VC dimension $VC(\mathcal{H})$ of a given hypothesis class $\mathcal{H}$ for the problem of binary classification is defined as the largest possible cardinality of some subset $\mathbf{X}' \subset \mathbf{X}$ for which there exists a hypothesis $h \in \mathcal{H}$ that perfectly classifies elements from $\mathbf{X}'$ whatever their classes are. More formally, we have*

$$VC(\mathcal{H}) = \max\{|\mathbf{X}'| : \forall y_i \in \{-1,+1\}^{|\mathbf{X}'|}, \exists h \in \mathcal{H} \textit{ so that } \forall \mathbf{x}_i \in \mathbf{X}', h(\mathbf{x}_i) = y_i\}.$$

As it follows from the definition, the VC dimension is the cardinality of the biggest subset of a given sample that can be subject to perfect classification provided by a hypothesis from $\mathcal{H}$ for all possible labelings of its observations. To illustrate it, we can consider a classical example given in Figure 1.3, where the hypothesis class $\mathcal{H}$ consists of half planes in $\mathbb{R}^d$. In this particular case, with $d = 2$, we can perfectly classify only $d + 1$ elements regardless of their labeling, as for the case with $d + 2$ points it will no longer be possible. It means that the VC dimension of the class of half-planes in $\mathbb{R}^d$ is equal to $d+1$. Note that the obtained result reveals that in this particular scenario, the VC dimension is equal to the number of parameters needed to define the function of the hypothesis plane. This, however, is not true in general as some classes may have an infinite VC dimension despite a finite number of parameters needed to define the hypothesis class. A common example used in the literature to show this is given by

$$\mathcal{H} = \{h_\theta(\mathbf{x}) : \mathbf{X} \to \{0, 1\} : h_\theta(\mathbf{x}) = \frac{1}{2}\sin(\theta\mathbf{x}), \theta \in \mathbb{R}\}.$$

It can be proven that the VC dimension of this class is infinite.

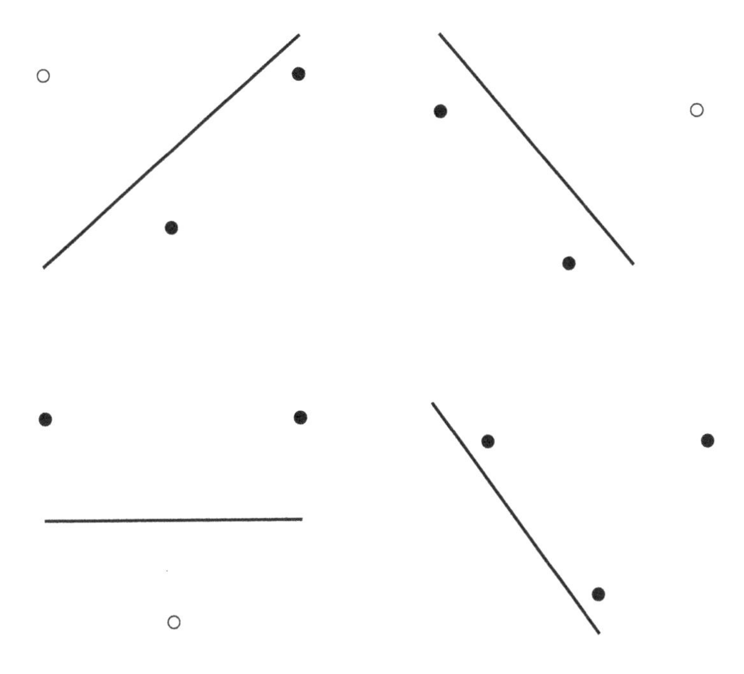

Figure 1.3. *Illustration of the idea behind the VC dimension. Here, half-planes in $\mathbb{R}^d$ with $d = 2$ can correctly classify at most three points for all possible labelings. The VC dimension here is $2 + 1$*

The following theorem uses the VC dimension of a hypothesis class to upper bound the gap between the true and the empirical error for a given loss function and a finite sample of size m.

THEOREM 1.2.– *Let* $\mathbf{X}$ *be an input space,* $Y = \{-1, +1\}$ *the output space and* $\mathcal{D}$ *their joint distribution. Let* S *be a finite sample of size* m *drawn i.i.d. from* $\mathcal{D}$ *and* $\mathcal{H} = \{h : X \to Y\}$ *be a hypothesis class of the VC dimension* $VC(\mathcal{H})$*. Then, for any* $\delta \in (0, 1]$ *with probability at least* $1-\delta$ *over the random choice of samples* $S \sim (\mathcal{D})^m$*, the following holds*

$$\sup_{h\in\mathcal{H}} \left|\mathrm{R}^{\ell}_{S}(h) - \mathrm{R}^{\ell}_{\mathcal{D}}(h)\right| \leq \sqrt{\frac{4}{m}\left(VC(\mathcal{H})\ln\frac{2em}{VC(\mathcal{H})} + \ln\frac{4}{\delta}\right)}.$$

As this bound is derived in terms of the supremum of the difference between two errors, we give the following corollary that states a similar result for every hypothesis $h \in \mathcal{H}$.

COROLLARY 1.1.– *Under the assumptions of theorem 1.2, for any* $\delta \in (0, 1]$ *with probability at least* $1 - \delta$ *over the random choice of the training sample* $S \sim (\mathcal{D})^m$*, the following holds*

$$\forall h \in \mathcal{H}, \quad \mathrm{R}^{\ell}_{\mathcal{D}}(h) \leq \mathrm{R}^{\ell}_{S}(h) + \sqrt{\frac{4}{m}\left(VC(\mathcal{H})\ln\frac{2em}{VC(\mathcal{H})} + \ln\frac{4}{\delta}\right)}.$$

We note that when $VC(\mathcal{H})$ is known, the right hand side of this inequality can be calculated explicitly. In general, this inequality indicates that for a certain confidence level given by $1-\delta$, the empirical risk of a hypothesis approaches its real value when m increases and this convergence is even faster for $\mathcal{H}$ with low VC dimension. Assuming that the VC dimension of $\mathcal{H}$ is finite and that h^*_S denotes a hypothesis minimizing the empirical risk, the law of big numbers implies convergence in probability of both $\mathrm{R}^{\ell}_{\mathcal{D}}(h)$ and $\mathrm{R}^{\ell}_{S}(h)$ to the optimal risk on $\mathcal{H}$. When this property is verified for a given learning algorithm, this latter is called *consistent*.

Even though the finiteness of the VC dimension ensures that the empirical risk will converge uniformly over the class to the true risk, in practice, the VC dimension based bounds present numerous inconveniences. First of all, deriving bounds for new learning algorithms using VC framework is usually quite laborious as for some hypothesis classes, the VC dimension is hard to calculate. On the other hand, the VC dimension was shown to be infinite for some popular machine learning methods (e.g. k-nearest neighbors), thus making the whole framework inapplicable to them. Furthermore, VC bounds are quite loose and thus present very small interest in practice: one has to estimate them based on samples consisting of a large number of data points in order to have a truthful approximation. VC bounds also do not take into account information about the used learning algorithm, nor do they depend on probability distribution of the data. In general, they can be seen as the worst case results of this type. In section 3.1, we present how this VC dimension theory has led to the first analysis of the domain adaptation problem considered in this book.

In the following section, we present generalization results based on the Rademacher complexity that allow us to overcome some drawbacks of VC bounds. Contrary to VC bounds that consider the worst-case labeling of $\mathcal{H}$, Rademacher-based bounds rely on the mean of all labelings. This has an advantage of incorporating the information about the data sample's probability distribution directly in the generalization bounds.

1.2.2. *Rademacher bounds*

Intuitively, Rademacher complexity measures the capacity of a given hypothesis class to resist against noise that may be present in data. This, in its turn, was shown to lead to more accurate bounds than those based on the VC dimension [KOL 99]. In order to present Rademacher bounds, we first give a definition of a Rademacher variable.

DEFINITION 1.4 (Rademacher variable).– *A random variable κ defined as*

$$\kappa = \begin{cases} 1, \text{ with probability } \frac{1}{2} \\ -1, \text{ otherwise,} \end{cases}$$

is called a Rademacher variable.

From this definition, a Rademacher variable defines random binary labeling as it takes values -1 and 1 with equal probability. We are now ready to introduce Rademacher complexity defined for an unlabeled sample of size m.

DEFINITION 1.5 (Rademacher complexity).– *For a given unlabeled sample $S = \{(\mathbf{x}_i)\}_{i=1}^m$ and a given hypothesis class $\mathcal{H}$, Rademacher complexity is defined as follows:*

$$\mathcal{R}_S(\mathcal{H}) = \mathop{\mathbf{E}}_{\boldsymbol{\kappa}} \left[\sup_{h \in \mathcal{H}} \frac{2}{m} \sum_{i=1}^{m} \kappa_i h(\mathbf{x}_i) \right],$$

where $\boldsymbol{\kappa}$ is a vector of m independent Rademacher variables. The Rademacher complexity for the whole hypothesis class is thus defined as the expected value of $\mathcal{R}_S(\mathcal{H})$

$$\mathcal{R}_m(\mathcal{H}) = \mathop{\mathbf{E}}_{S \sim (\mathcal{D})^m} \mathcal{R}_S(\mathcal{H}).$$

In this definition, $\mathcal{R}_S(\mathcal{H})$ encodes the complexity of a given hypothesis class $\mathcal{H}$ based on the observed sample S, while $\mathcal{R}_m(\mathcal{H})$ stands for the expected value of this complexity over all possible samples that were drawn from some joint probability distribution. Contrary to the VC dimension, this complexity measure is defined in terms of the expected value over all labelings and not only the worst one. The following theorem presents the Rademacher-based bounds based on empirical and true Rademacher complexities [KOL 99, BAR 02].

THEOREM 1.3.– *Let $S = \{(\mathbf{x}_i, y_i)\}_{i=1}^m$ be a finite sample of m examples drawn i.i.d. from $\mathcal{D}$, and $\mathcal{H} = \{h : \mathbf{X} \to Y\}$ be a hypothesis class. Then, for any $\delta \in (0,1]$ with probability at least $1-\delta$ over the choice of the sample $S \sim (\mathcal{D})^m$, the following holds*

$$\forall h \in \mathcal{H}, \quad \mathrm{R}^{\ell}_{\mathcal{D}}(h) \leq \mathrm{R}^{\ell}_{S}(h) + \mathcal{R}_m(\mathcal{H}) + \sqrt{\frac{\ln \frac{1}{\delta}}{2m}}, \qquad [1.2]$$

$$\mathrm{R}^{\ell}_{\mathcal{D}}(h) \leq \mathrm{R}^{\ell}_{S}(h) + \mathcal{R}_S(\mathcal{H}) + 3\sqrt{\frac{\ln \frac{2}{\delta}}{2m}}. \qquad [1.3]$$

Equation [1.3] of this theorem includes the empirical Rademacher complexity $\mathcal{R}_S(\mathcal{H})$: a term that can be explicitly calculated using the sample S unlike VC dimension-based results. This important difference between Rademacher complexity and the VC dimension highlights the advantages that Rademacher generalization bounds offer in comparison to the statistical learning bounds of VC theory. These bounds also present an example of what is usually called data-dependent bounds, as their calculation relies on the knowledge about the considered data samples and not on that of the topological structure of the hypothesis space.

To calculate them, we note that the Rademacher average can be rewritten as follows:

$$\begin{aligned}
\frac{1}{2}\mathcal{R}_S(\mathcal{H}) &= \frac{1}{2}\hat{A}\mathop{\mathbf{E}}_{\boldsymbol{\kappa}}\left[\sup_{h\in\mathcal{H}} \frac{2}{m}\sum_{i=1}^{m} \kappa_i h(\mathbf{x}_i)\right] \\
&= \frac{1}{2} + \mathop{\mathbf{E}}_{\boldsymbol{\kappa}}\left[\sup_{h\in\mathcal{H}} \frac{2}{m}\sum_{i=1}^{m} \frac{-1-\kappa_i h(x_i)}{2}\right] \\
&= \frac{1}{2} - \mathop{\mathbf{E}}_{\boldsymbol{\kappa}}\left[\inf_{h\in\mathcal{H}} -\frac{2}{m}\sum_{i=1}^{m} \frac{1-\kappa_i h(\mathbf{x}_i)}{2}\right] \\
&= \frac{1}{2} - \inf_{h\in\mathcal{H}} \mathrm{R}^{\ell}_{S}(h,\kappa).
\end{aligned}$$

From the last expression, we may note that computing the Rademacher average for a given sample S and fixed variables κ_i is equal to computing the empirical risk minimizer from the hypothesis class $\mathcal{H}$. One may note that as $\boldsymbol{\kappa}$ essentially represents random binary variables, the Rademacher average quantifies the capacity of a given class to fit the random noise.

While the Rademacher average is quite different from the concept of the VC dimension, the two can be related using the following inequality:

$$\mathcal{R}_m(\mathcal{H}) \leq 2\sqrt{\frac{4}{m}\left(VC(\mathcal{H})\ln\frac{2em}{VC(\mathcal{H})} + \ln\frac{4}{\delta}\right)}.$$

This result closely relates the original VC dimension-based bounds presented above to the Rademacher complexity-based theory.

Finally, we note that similar to the VC dimension-based bounds, the domain adaptation problem has also been studied using the tools from the theory of Rademacher complexity. We present this analysis in section 3.5. The following section introduces generalization bounds that are expressed as expectation over the hypothesis space.

1.2.3. *PAC–Bayesian bounds*

The PAC–Bayesian approach [MCA 99] provides PAC generalization bounds for hypothesis expressed as a weighted majority vote[1] over a set of functions that can be the hypothesis space $\mathcal{H}$. In this section, we recall the general PAC–Bayesian generalization bound as presented by Germain *et al.* [GER 15b] in the setting of binary classification where $Y = \{-1, 1\}$ with the $0-1$ loss or the linear loss. To derive such a generalization bound, one assumes a prior distribution π over $\mathcal{H}$, which models an *a priori* belief on the hypothesis from $\mathcal{H}$ before the observation of the training sample $S \sim (\mathcal{D})^m$. Given the training sample S, the learner aims at finding a posterior distribution ρ over $\mathcal{H}$ that leads to a well-performing ρ-weighted majority vote[2] $B_\rho(\mathbf{x})$ defined as:

$$B_\rho(\mathbf{x}) = \text{sign}\left[\mathop{\mathbf{E}}_{h\sim\rho} h(\mathbf{x})\right].$$

In other words, rather than finding the best hypothesis from $\mathcal{H}$, we want to learn ρ over $\mathcal{H}$ such that it minimizes the true risk $\mathrm{R}_\mathcal{D}(B_\rho)$ of the ρ-weighted majority vote. However, PAC–Bayesian generalization bounds do not directly focus on the risk of the deterministic ρ-weighted majority vote $B_\rho(\cdot)$. Instead, it gives an upper bound over

1 Note that the majority vote setting is not too restrictive since many machine learning approaches can be considered as majority vote learning, notably ensemble methods [DIE 00, RE 12].

2 In the PAC–Bayesian literature, it is sometimes called the Bayes classifier.

the expectation over ρ of all the individual hypothesis true risk called the *Gibbs risk*[3]: $\mathbf{E}_{h\sim\rho}\,\mathrm{R}^{\ell}_{\mathcal{D}}(h)$. PAC–Bayesian generalization bounds are then expressed in terms of expectation over all the hypothesis space (instead of focusing on a single hypothesis). An important behavior of the Gibbs risk is that it is closely related to the deterministic ρ-weighted majority vote. Indeed, if $B_\rho(\cdot)$ misclassifies $\mathbf{x} \in \mathbf{X}$, then at least half of the classifiers (under measure ρ) make a prediction error on $\mathbf{x}$. Therefore, we have

$$\mathrm{R}^{\ell}_{\mathcal{D}}(B_\rho) \leq 2 \mathop{\mathbf{E}}_{h\sim\rho} \mathrm{R}^{\ell}_{\mathcal{D}}(h). \qquad [1.4]$$

Thus, an upper bound on $\mathop{\mathbf{E}}_{h\sim\rho} \mathrm{R}^{\ell}_{\mathcal{D}}(h)$ gives rise to an upper bound on $\mathrm{R}^{\ell}_{\mathcal{D}}(B_\rho)$. Note that other tighter relations exist [LAN 02, LAC 06, GER 15b].

Note that PAC–Bayesian generalization bounds do not directly take into account the complexity of the hypothesis class $\mathcal{H}$, but measure the deviation between the prior distribution π and the posterior distribution ρ on $\mathcal{H}$ through the Kullback–Leibler divergence[4]:

$$\mathrm{KL}(\rho|\pi) = \mathop{\mathbf{E}}_{h\sim\rho} \ln \frac{\rho(h)}{\pi(h)}. \qquad [1.5]$$

The following theorem is a general PAC–Bayesian theorem that takes the form of an upper bound on the deviation, according to a convex function $D : [0,1] \times [0,1] \to \mathbb{R}$, between the true and empirical Gibbs risks.

THEOREM 1.4 ([GER 09, GER 15b]).– *For any distribution $\mathcal{D}$ on $\mathbf{X} \times Y$, for any hypothesis class $\mathcal{H}$, for any prior distribution π on $Hcal$, for any $\delta \in (0,1]$ and for any convex function $D : [0,1] \times [0,1] \to \mathbb{R}$, with a probability at least $1-\delta$ over the random choice of $S \sim (\mathcal{D})^m$, we have, for all posterior distribution ρ on $\mathcal{H}$,*

$$D\left(\mathop{\mathbf{E}}_{h\sim\rho} \mathrm{R}^{\ell}_{S}(h), \mathop{\mathbf{E}}_{h\sim\rho} \mathrm{R}^{\ell}_{\mathcal{D}}(h)\right)$$
$$\leq \frac{1}{m}\left[\mathrm{KL}(\rho|\pi) + \ln\left(\frac{1}{\delta} \mathop{\mathbf{E}}_{S\sim(\mathcal{D})^m} \mathop{\mathbf{E}}_{h\sim\pi} e^{m\,D\left(\mathrm{R}^{\ell}_{S}(h), \mathrm{R}^{\ell}_{\mathcal{D}}(h)\right)}\right)\right].$$

By upper bounding $\mathop{\mathbf{E}}_{S\sim(\mathcal{D})^m} \mathop{\mathbf{E}}_{h\sim\pi} e^{m\,D(\mathrm{R}^{\ell}_{S}(h))}$ and by selecting a well-suited deviation function D, we can retrieve the classical versions of the PAC–Bayesian theorem (i.e. [MCA 99, SEE 02, CAT 07]).

3 The Gibbs risk is associated with a stochastic classifier called the Gibbs classifier, which predicts the label of an example $\mathbf{x}$ by drawing h from $\mathcal{H}$ according to the posterior distribution ρ and predicts $h(\mathbf{x})$.

4 It is worth noting that the KL-divergence term can "disappear" in some situations. See [GER 15b] for more details.

For the sake of simplicity, we instantiate here theorem 1.4 to obtain the version of Catoni [CAT 07] that will be useful for domain adaptation in Chapter 6. To do so, we put $D(a,b) = \ln\left(\frac{1}{b[1-\exp(-\omega)]}\right) - \omega\, a$, with $\omega > 0$. Note that the theorem below is expressed in the simplified form suggested by Germain *et al.* [GER 09].

THEOREM 1.5 ([CAT 07]).– *For any distribution $\mathcal{D}$ on $\mathbf{X}\times Y$, for any set of voters $\mathcal{H}$, for any prior distribution π on Hcal, for any $\delta \in (0,1]$, and any real number $\omega > 0$, with a probability at least $1-\delta$ over the random choice of the sample $S \sim (\mathcal{D})^m$, we have, for all posterior distribution ρ on $\mathcal{H}$,*

$$\mathop{\mathbf{E}}_{h\sim\rho} \mathrm{R}^{\ell}_{\mathcal{D}}(h) \le \frac{\omega}{1-e^{-\omega}}\left[\mathop{\mathbf{E}}_{h\sim\rho} \mathrm{R}^{\ell}_{S}(h) + \frac{\mathrm{KL}(\rho|\pi)) + \ln\frac{1}{\delta}}{m\times\omega}\right]. \qquad [1.6]$$

Similarly to the classical PAC–Bayesian theorems, the theorem above suggests that minimizing the expected risk can be done by minimizing the trade-off between the empirical risk $\mathrm{R}^{\ell}_{S}(G_\rho)$ and KL-divergence minimization $\mathrm{KL}(\rho|\pi)$ (roughly speaking the complexity term).

Lastly, it is important to remark that under the assumption that $\mathcal{H}$ is finite, if we suppose that the prior distribution π is the uniform distribution, and that the posterior distribution ρ is defined such that $\rho(h_S) = 1$ for some $h_S \in \mathcal{H}$ and $\rho(h) = 0$ for all other $h \in \mathcal{H}$, with $D(a,b) = (a-b)^2$, then we have

$$\mathrm{R}^{\ell}_{\mathcal{D}}(h_S) \le \mathrm{R}^{\ell}_{S}(h_S) + \sqrt{\frac{\ln(|\mathcal{H}|) + \ln(m/\delta)}{2(m-1)}}.$$

This bound brings similar ideas than usual VC dimension-based bounds (see corollary 1.1).

Many PAC–Bayesian studies have been conducted to characterize the error of majority votes [CAT 07, SEE 02, LAN 02, GER 15b] as well as to derive PAC–Bayesian theoretically grounded learning algorithms, namely in the supervised learning setting (e.g. [GER 09, PAR 12, ALQ 16, ROY 16]) and the domain adaptation setting (e.g. [GER 16, GER 13], see Chapter 6).

In the following sections, we present a couple of other data-dependent bounds that allow us to exclude the complexity of the hypothesis class from the bounds and to link directly the generalization performance to the properties of a given learning algorithm.

1.2.4. *Uniform stability*

From the previous sections, we have already seen that there may be different strategies used to find the best hypothesis from a given hypothesis class. These strategies correspond to different learning algorithms that explore the hypothesis space by solving an appropriate optimization problem. As the complexity of the hypothesis class in this case depends directly on the properties of a learning algorithm, it may be desirable to have the generalization bounds that manifest this relationship explicitly. Bousquet and Elisseeff [BOU 02] introduced the generalization bounds that provide a solution to this problem based on the notion of uniform stability of a learning algorithm. We now give its definition.

DEFINITION 1.6 (Uniform stability).– *An algorithm $\mathcal{A}$ has uniform stability β with respect to the loss function ℓ if the following holds*

$$\forall S \in \{\mathbf{X} \times Y\}^m, \forall i \in \{1, \ldots, m\}, \sup_{(\mathbf{x},y)\in S} |\ell(h_S(\mathbf{x}), y) - \ell(h_{S^{\setminus i}}(\mathbf{x}), y)| \leq \beta,$$

where hypothesis h_S is learned on the sample S while $h_{S^{\setminus i}}$ is obtained on the sample where the ith *observation was deleted.*

The intuition behind this definition is to say that an algorithm that is expected to generalize well should be robust to the small perturbations in the training sample. Consequently, stable algorithms should have their empirical error remain close to their generalization error. This idea is confirmed by the following theorem.

THEOREM 1.6.– *Let $\mathcal{A}$ be an algorithm with uniform stability β with respect to a loss function ℓ such that $0 \leq \ell(h_S(\mathbf{x}, y) \leq M$, for all $(\mathbf{x}, y) \in (\mathbf{X} \times Y)$ and all sets S. Then, for any $m \geq 1$, and any $\delta \in (0, 1]$, the following bound holds with probability at least $1 - \delta$ over the random choice of the sample S,*

$$\mathrm{R}^{\ell}_{\mathcal{D}}(h) \leq \mathrm{R}^{\ell}_{S}(h) + 2\beta + (4m\beta + M)\sqrt{\frac{\ln \frac{1}{\delta}}{2m}}.$$

This theorem states that an algorithm with uniform stability β generalizes well when β scales as $\frac{1}{m}$. When this condition is verified for a given learning algorithm, this latter is called *uniformly stable*. We also note that the β term usually depends on the loss function and the regularization type used by a learning algorithm and many famous machine learning algorithms were shown to verify the uniform stability property. The important advantage of stability bounds lies in their independence of the hypothesis class complexity for a given learning algorithm: even when it is hard to estimate or infinite, the stability bounds can still be calculated explicitly.

As we stated above, it has been shown that many famous supervised learning algorithms are uniformly stable and thus the proposed bounds can be applied to them. This type of results can also be extended to the domain adaptation problem considered in this book. We present theoretical results derived based on this theory for the domain adaptation problem in Chapter 7.

While stability theory addresses many issues related to VC and Rademacher-based bounds, there exists a couple of notorious exceptions of algorithms found by Xu and Mannor [XU 10] that do not verify the uniform stability property. The latter includes, for instance, learning models with L_1-norm penalty popularly used to enforce sparsity of the obtained solution. In order to propose an even more general theoretical framework that would include these sparsity inducing models, Xu and Mannor [XU 10] introduced the notion of algorithmic robustness presented in the following.

1.2.5. *Algorithmic robustness*

The main underlying idea of algorithmic robustness [XU 10] is to say that a robust algorithm should have similar performance in terms of the classification error on a testing and training samples that are close. The measure of similarity used to define if two samples are close or not relies on partitioning the product space $\mathbf{X} \times Y$ in a way that put two similar points of the same class to the same partition. This partition is further defined using the notion of covering numbers [KOL 59] introduced below.

DEFINITION 1.7 (Covering number).– *Let (Z, ϱ) denote a metric space with metric $\varrho(\cdot)$ defined on Z. For $Z' \subset Z$, we say that $\hat{Z}'$ is a γ covering of Z'; for any element $t \in Z'$, there is an element $\hat{t} \in \hat{Z}'$ such that $\varrho(t, \hat{t}) \leq \gamma$. Then the γ-covering number of Z' is expressed as*

$$N(\gamma, Z', \varrho) = \min\left\{\left|\hat{Z}'\right| : \hat{Z}' \textit{ is a } \gamma\textit{-covering of } Z'\right\}.$$

In the case where $\mathbf{X}$ is a compact space, its covering number $N(\gamma, \mathbf{X}, \varrho)$ is finite. Furthermore, for the product space $\mathbf{X} \times Y$, the number of γ-covering is also finite and equals $|Y|\, N(\gamma, \mathbf{X}, \varrho)$. As previously explained, the above partitioning ensures that two points from the same subset are from the same class and are close to each other with respect to metric ϱ. Bearing this in mind, the algorithmic robustness is defined as follows.

DEFINITION 1.8.– *Let S be a training sample of size m where each example is drawn from the joint distribution $\mathcal{D}$ on $\mathbf{X} \times Y$. An algorithm $\mathcal{A}$ is $(M, \epsilon(S))$-robust on $\mathcal{D}$ with respect to a loss function ℓ for $M \in \mathbb{N}$ and $\epsilon : (\mathbf{X} \times Y)^m \to \mathbb{R}$ if $\mathbf{X} \times Y$ can be partitioned into M disjoint subsets denoted by $\{Z_j\}_{j=1}^{M}$ so that for all $(\mathbf{x}, y) \in S$, $(\mathbf{x}', y')$ drawn from $\mathcal{D}$ and $j \in \{1, \ldots, M\}$, we have*

$$((\mathbf{x}, y), (\mathbf{x}', y')) \in Z_j^2 \quad \longrightarrow \quad |\ell(h_S(\mathbf{x}), y) - \ell(h_S(\mathbf{x}'), y')| \leq \epsilon(S),$$

where h_S is a hypothesis learned by $\mathcal{A}$ on S.

We are now ready to present the generalization guarantees that characterize robust algorithms verifying the definition presented above.

THEOREM 1.7.– *Let S be a finite sample of size m drawn i.i.d. from $\mathcal{D}$, $\mathcal{A}$ be $(M, \epsilon(S))$-robust on $\mathcal{D}$ with respect to a loss function $\ell(\cdot,\cdot)$ such that $0 \leq \ell(h_S(\mathbf{x}, y) \leq M_\ell$, for all $(\mathbf{x}, y) \in (\mathbf{X} \times Y)$. Then, for any $\delta \in (0, 1]$, the following bound holds with probability at least $1 - \delta$ over the random draw of the sample $S \sim (\mathcal{D})^m$,*

$$\mathrm{R}^\ell_{\mathcal{D}}(h) \leq \mathrm{R}^\ell_S(h) + \epsilon(S) + M_\ell \sqrt{\frac{2M \ln 2 + 2 \ln \frac{1}{\delta}}{m}},$$

where h_S is a hypothesis learned by $\mathcal{A}$ on S.

As $\epsilon(S)$ depends on the γ-covering of $\mathbf{X} \times Y$ and its size M, it naturally implies a trade-off between M and $\epsilon(S)$. Similar to the bounds based on the uniform stability, the result of this theorem does not depend on the complexity of the hypothesis class and thus its right-hand side can be calculated even if the latter is not computable or infinite. An important difference between them lies in the definition of algorithmic robustness that focuses on measuring the divergence between the costs associated with two similar points, assuming that the learned hypothesis function should be locally consistent. Uniform stability, in its turn, explores the variation in the cost provoked by slight perturbations of the training sample and thus assumes that the learned hypothesis does not change much. In Chapter 7, we present an extension of this notion of robustness proposed by Mansour and Schain [MAN 14] for the domain adaptation problem.

1.3. Summary

In this chapter, we formalized the classical supervised learning problem, the assumptions that are usually imposed in this setting that relate the observed (training) and unseen (test) samples. We explained different risk minimization strategies that can be used to produce a "good" hypothesis for a problem at hand. The motivation to have these different strategies was illustrated by the "no-free lunch" theorem that allowed us to explicitly show the non-existence of a universal learner that is good for any given task and the need for prior knowledge that guides a learning algorithm. This general introduction was followed by a brief overview of different generalization guarantees and a discussion regarding their advantages and inconveniences. To sum up, the presented generalization bounds prove that the convergence of the empirical risk to the true risk with the increasing number of available samples may depend on several terms:

1) the capacity of a hypothesis class in solving a given classification problem (i.e. the VC dimension and Rademacher averages based bounds);

2) the proximity of the *a priori* knowledge regarding the form of the hypothesis (i.e. PAC–Bayesian bounds);

3) the properties of the used learning algorithm (i.e. uniform stability and robustness-based bounds).

In general, the theoretical results presented in this chapter in all scenarios share an important common feature, that is to explicitly assume that training and test samples are generated by the same (unknown) probability distribution. This seems reasonable as, indeed, one often wants to learn a model in order to make predictions on a data sample that is somehow very "close" to the data that the model was trained on. Obviously, in many real-world applications, this proximity is rather approximate as training data and newly collected test data may often exhibit different statistical characteristics. The extent to which these differences may violate the assumption regarding the same distribution of training and test data may more or less position the learning algorithm out of scope of the standard supervised learning theory. In this case, a new learning scenario that reflects this particular situation arises. It is known as domain adaptation and is the main subject of this book.

2

Domain Adaptation Problem

In general, the concept of learning can be divided into two broad settings based on the presence or absence of a teacher or relevant prior knowledge that can replace it. In the first case, learning mainly concerns the acquisition of knowledge from scratch, for which full permanent supervision is obviously necessary and cannot be replaced. This particular case is reminiscent of a human being who learns without any prior knowledge and therefore needs a teacher to provide them with all the necessary information. This information usually takes the form of oral instructions, annotations or written labels. Obviously, this type of learning takes place mainly among very young children who learn by following direct instructions from their parents or other family members. From a certain age, the human being generally acquires a sufficient basis to learn new things without supervision or under partial supervision. This type of learning often defines a person's general intelligence and intuition because it reflects their ability to generalize to different applications and thus their ability to operate at a higher abstract level.

As machine learning is essentially inspired by the phenomenon of human learning, it is not surprising that the same hierarchy of learning algorithms can be found in its literature too. Indeed, it can be noted that a large majority of traditional methods use learning in a strictly supervised way, relying on huge amounts of annotated data. This type of approach, generally referred to as *supervised learning*, was the first to be proposed by scientists in the machine learning field and remains among the most popular nowadays because of the abundant amount of collected labeled data allowing its widespread deployment.

More recent methods address a more challenging learning setting with only partially labeled data under the assumption that these data are sufficient to extract important patterns that will allow one to generalize on unseen and unlabeled points. This learning paradigm, called *semi-supervised learning*, is often used in applications where data are abundant but expensive to label, which means that not only labeled data but also unlabeled data must be considered to learn effectively. Finally, the last major family of traditional machine learning algorithms is *unsupervised learning*, whose objective is to group data into homogeneous sets by ensuring a maximum degree of separability between these sets. This learning paradigm can be considered as an exploratory tool to obtain a first overview of a large amount of unlabeled data and thus to find patterns that are potentially different from those provided by manual labeling.

In addition to these three learning paradigms, an even more ambitious step undertaken by scientists in recent years is the development of algorithms that allow us to learn and extrapolate knowledge across data coming from different domains. The idea behind this learning paradigm, usually referred to as *transfer learning*, is inspired by the human being's ability to learn independently with minimal or no supervision based on previously acquired knowledge. It is not surprising that this concept was not invented in the machine learning community in the proper sense of the term, since the term "transfer of learning" had been used long before the construction of the first computer and is found in the psychology field papers from the early 20th Century.

2.1. Notations and preliminaries

In the previous chapter, we described some major families of generalization bounds proved for traditional supervised learning. As we have shown, these generalization bounds allow one to upper bound the true risk of a hypothesis on a particular data distribution by its empirical risk evaluated on the available learning sample where each data point is drawn from the same probability distribution plus terms related to the complexity of the hypothesis set or the properties of the algorithm used. Despite the usefulness of the presented results, they only hold under the strong assumption stating that the training and test data have to be drawn from the same probability distribution. This assumption, however, is often too restrictive to hold in practice as in many real-world applications, a hypothesis is learned and deployed in environments that differ and exhibit an important shift. The following examples can be used to illustrate this situation.

Spam filtering problem. Consider a situation where a spam filter is learned using a classification algorithm for a corporate mailbox of a given user. In this case, the vast majority of e-mails analyzed by the algorithm are likely to be of a professional character with very few of them being related to the private life of the person considered. Imagine further a situation where this same user installs a mailbox software on the personal computer and imports the settings of its corporate mailbox hoping that it will work equally well on it. This, however, is not likely to be the case as many personal e-mails may seem like spam to an algorithm learned purely on professional communications due to the differences in their content and attached files as well as the non-uniformity of e-mail addresses.

Fraud detection. Imagine that a certain company provides software that detects fraud in banking transactions based on data that it thoroughly collected over the last 5 years. Then, a new payment method gets introduced by the government relying on a different security protocol. If the considered company decides to apply the same algorithm to the newly collected data in order to detect fraud, their performance is likely to degrade due to the changing behavior of the frauds that were obliged to adapt to new security protocols in order to bypass them.

Species classification. In many oceanographic studies, one relies on the video coverage of a certain sea area in order to recognize species of the marine habitat. For instance, in the Mediterranean Sea and in the Indian Ocean, the species of fish that can be found on the recorded videos are likely to belong to the same family of fishes, even though their actual appearance may be quite different due to different climate and evolutionary backgrounds. In this case, the learning algorithm learned on the video coverage of the Mediterranean Sea will most likely fail to provide a correct classification of species in the Indian Ocean if it is not specifically adapted by an expert.

For all applications considered above, it may be desirable to find a learning paradigm that can remain robust to a changing environment. To this end, the above-mentioned paradigm should be able to adapt to a new problem at hand by drawing parallels and exploiting the knowledge from the domain where it was learned in the first place. In response to this problem, the quest for new algorithms able to learn on a training sample and then have a good performance on a test sample coming from a different but related probability distribution gave rise to a new learning paradigm called *transfer learning*. Its definition is given as follows.

DEFINITION 2.1 (Transfer learning).– *We consider a source data distribution $\mathcal{S}$ called the source domain, and a target data distribution $\mathcal{T}$ called the target domain. Let $\mathbf{X}_\mathcal{S} \times Y_\mathcal{S}$ be the source input and output spaces associated with $\mathcal{S}$, and $\mathbf{X}_\mathcal{T} \times Y_\mathcal{T}$ be the target input and output spaces associated $\mathcal{T}$. We denote by $\mathcal{S}_\mathbf{X}$ and $\mathcal{T}_\mathbf{X}$ the marginal distributions of $\mathbf{X}_\mathcal{S}$ and $\mathbf{X}_\mathcal{T}$, and by $t_\mathcal{S}$ and $t_\mathcal{T}$ the source and target learning tasks depending on $Y_\mathcal{S}$ and $Y_\mathcal{T}$, respectively. Then, transfer learning aims to improve the learning of the target predictive function $f_\mathcal{T} : \mathbf{X}_\mathcal{T} \to Y_\mathcal{T}$ for $t_\mathcal{T}$ using knowledge gained from $\mathcal{S}$ where $\mathcal{S} \neq \mathcal{T}$.*

Note that the condition $\mathcal{S} \neq \mathcal{T}$ implies either $\mathcal{S}_\mathbf{X} \neq \mathcal{T}_\mathbf{X}$ (i.e. $\mathbf{X}_\mathcal{S} \neq \mathbf{X}_\mathcal{T}$ or $\mathcal{S}_\mathbf{X}(\mathbf{X}) \neq \mathcal{T}_\mathbf{X}(\mathbf{X})$) or $t_\mathcal{S} \neq t_\mathcal{T}$ (i.e. $Y_\mathcal{S} \neq Y_\mathcal{T}$ or $\mathcal{S}(Y|\mathbf{X}) \neq \mathcal{T}(Y|\mathbf{X})$).

In transfer learning, one often distinguishes three possible learning settings based on these different relations (illustrated in Figure 2.1):

1) Inductive transfer learning where $\mathcal{S}_\mathbf{X} = \mathcal{T}_\mathbf{X}$ and $t_\mathcal{S} \neq t_\mathcal{T}$:
For example, $\mathcal{S}_\mathbf{X}$ and $\mathcal{T}_\mathbf{X}$ are the distributions of the data collected from the mailbox

of one particular user, with $t_{\mathcal{S}}$ *being the task of detecting spam, while* $t_{\mathcal{T}}$ *being the task of detecting a hoax.*

2) Transductive transfer learning where $\mathcal{S}_{\mathbf{X}} \neq \mathcal{T}_{\mathbf{X}}$ but $t_{\mathcal{S}} = t_{\mathcal{T}}$:
For example, in the spam filtering problem, $\mathcal{S}_{\mathbf{X}}$ *is the distribution of the data collected for one user,* $\mathcal{T}_{\mathbf{X}}$ *is the distribution of data of another user and* $t_{\mathcal{S}}$ *and* $t_{\mathcal{T}}$ *are both the task of detecting spam.*

3) Unsupervised transfer learning where $t_{\mathcal{S}} \neq t_{\mathcal{T}}$ and $\mathcal{S}_{\mathbf{X}} \neq \mathcal{T}_{\mathbf{X}}$:
For example, $\mathcal{S}_{\mathbf{X}}$ *generates the data collected from one user and* $\mathcal{T}_{\mathbf{X}}$ *generates the content of web pages collected from the web with* $t_{\mathcal{S}}$ *consisting of detecting spam, while* $t_{\mathcal{T}}$ *is to detect hoaxes.*

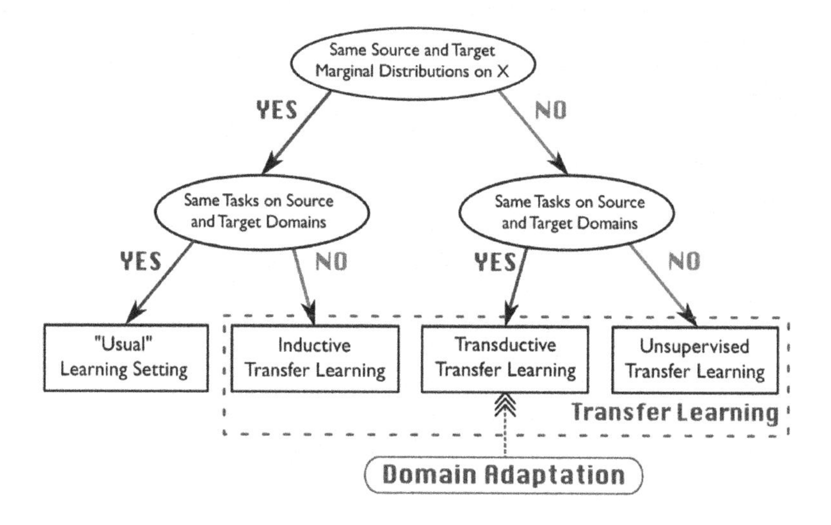

Figure 2.1. *Distinction between usual supervised learning setting and transfer learning, and positioning of domain adaptation. For a color version of this figure, see www.iste.co.uk/redko/domain.zip*

Arguably, the vast majority of situations where transfer learning proves to be the most needed falls into the second category. This latter has the name of *domain adaptation*, where we suppose that the source and the target tasks are the same, but where we have a source data set with an abundant amount of labeled observations and a target one with no (or little) labeled instances.

2.2. Overview of the adaptation scenarios

In general, transfer learning is related to any learning problem where the joint distributions of training and test samples $\mathcal{S}$ and $\mathcal{T}$ defined over a product space of features and labels $\mathbf{X} \times Y$ are different. From the probabilistic point of view, this inequality can be due to several reasons that differ in the nature of the causal link between the instances and labels. To this end, we define:

1) $\mathbf{X} \rightarrow Y$ problems, where the class label is causally determined by the values of the instances. This scenario can be related to the vast majority of machine learning tasks where the description of the object from $\mathbf{X}$ defines its label from Y. For instance, in image classification, the descriptors of an image define the category that it belongs to. In this case, we can write the decomposition of a given joint distribution $\mathcal{P}$ as

$$P(\mathbf{X}, Y) = P(\mathbf{X})P(Y|\mathbf{X}).$$

2) $Y \rightarrow \mathbf{X}$ problems, where the class label causally determines the values of the instances. This particular case can often be encountered in medical applications where symptoms are caused by the disease that one tries to predict. In this case, we can write the decomposition of a given joint distribution $\mathcal{P}$ as

$$P(\mathbf{X}, Y) = P(Y)P(\mathbf{X}|Y).$$

In what follows, we use these factorizations to give a comprehensive presentation of different types of shift that exist in the literature. In this characterization, we build upon the work of Moreno-Torres *et al.* [MOR 12a], where the authors studied and systematized different terms describing data set shift in several large categories.

2.2.1. *Covariate shift*

To proceed, we first consider the $\mathbf{X} \rightarrow Y$ problem where the inequality between $\mathcal{S}$ and $\mathcal{T}$ can be due to:

- $\mathcal{S}_{\mathbf{X}}(\mathbf{X}) \neq \mathcal{T}_{\mathbf{X}}(\mathbf{X})$ while $\mathcal{S}(Y|\mathbf{X}) = \mathcal{T}(Y|\mathbf{X})$;
- $\mathcal{S}_{\mathbf{X}}(\mathbf{X}) = \mathcal{T}_{\mathbf{X}}(\mathbf{X})$ while $\mathcal{S}(Y|\mathbf{X}) \neq \mathcal{T}(Y|\mathbf{X})$.

The first of the two considered cases corresponds to a problem of *covariate shift* [SHI 00]. In covariate shift, we assume that the marginal distributions of source and target data change while the predictive dependency remains unchanged. This situation is illustrated in Figure 2.2. A good example of a real-world covariate shift problem is given by the famous Office/Caltech [SAE 10, GOP 11] benchmark data set consisting of four domains:

1) Amazon: images from online merchants (958 images with 800 features from 10 classes);

2) Webcam: set of low-quality images by a web camera (295 images with 800 features from 10 classes);

3) DSLR: high-quality images by a digital SLR camera (157 images with 800 features from 10 classes);

4) Caltech: famous data set for object recognition (1,123 images with 800 features from 10 classes).

Several images from this data set are represented in Figure 2.3. From these images, we can see that the objects from the same classes drawn from different domains are all subject to high variability in terms of the image quality and its representation, which means that a classifier learned on the sample of one of these domains will likely fail to perform well on the other domain.

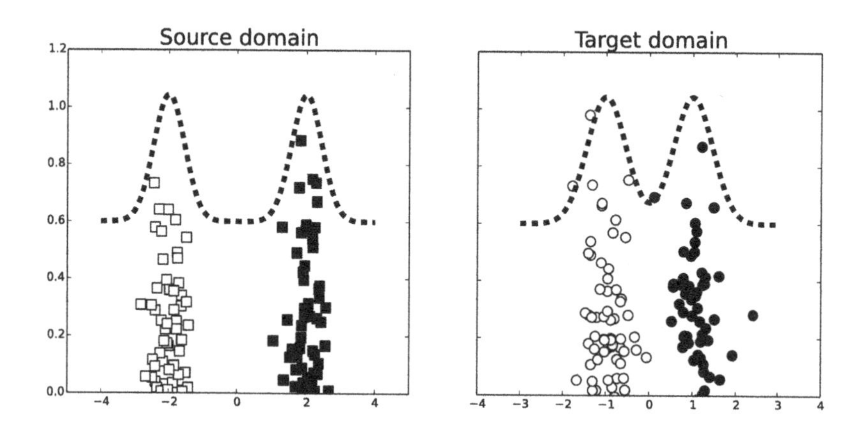

Figure 2.2. *Illustration of covariate shift. Here, marginal distributions of source domains (left) and target domains (right) are different while the conditional distributions remain the same*

Following the arguments from [SHI 00], the covariate shift problem may be solved because of a reweighting scheme in the following way:

$$
\begin{aligned}
\mathrm{R}^{\ell}_{\mathcal{T}}(h) &= \mathop{\mathbf{E}}_{(\mathbf{x},y)\sim\mathcal{T}} \ell(h(\mathbf{x}),y) \\
&= \mathop{\mathbf{E}}_{(\mathbf{x},y)\sim\mathcal{T}} \frac{\mathcal{S}(\mathbf{x},y)}{\mathcal{S}(\mathbf{x},y)} \ell(h(\mathbf{x}),y) \\
&= \sum_{(\mathbf{x},y)\,\in\,\mathbf{X}\times Y} \mathcal{T}(\mathbf{x},y) \frac{\mathcal{S}(\mathbf{x},y)}{\mathcal{S}(\mathbf{x},y)} \ell(h(\mathbf{x}),y) \\
&= \mathop{\mathbf{E}}_{(\mathbf{x},y)\sim\mathcal{S}} \frac{\mathcal{T}(\mathbf{x},y)}{\mathcal{S}(\mathbf{x},y)} \ell(h(\mathbf{x}),y).
\end{aligned}
$$

Figure 2.3. *Examples of keyboard and backpack images from Amazon, Caltech, DSLR and Webcam data sets from left to right. One can observe that images from the class differ drastically from one domain to another*

Now, using the assumption $\mathcal{S}(y|\mathbf{x}) = \mathcal{T}(y|\mathbf{x})$ gives

$$\begin{aligned} \mathrm{R}_{\mathcal{T}} &= \mathop{\mathbb{E}}_{(\mathbf{x},y)\sim\mathcal{S}} \frac{\mathcal{T}_{\mathbf{X}}(\mathbf{x})\mathcal{T}(y|\mathbf{x})}{\mathcal{S}_{\mathbf{X}}(\mathbf{x})\mathcal{S}(y|\mathbf{x})} \ell(h(\mathbf{x}), y) \\ &= \mathop{\mathbb{E}}_{(\mathbf{x},y)\sim\mathcal{S}} \frac{\mathcal{T}_{\mathbf{X}}(\mathbf{x})}{\mathcal{S}_{\mathbf{X}}(\mathbf{x})} \ell(h(\mathbf{x}), y). \end{aligned}$$

The natural idea then is to reweight labeled source data according to an estimate of $\frac{\mathcal{T}_{\mathbf{X}}(\mathbf{x})}{\mathcal{S}_{\mathbf{X}}(\mathbf{x})}$. This reweighting scheme leads to a large family of instance-reweighting methods that were proposed to tackle the covariate shift problem. In its essence, instance-reweighting methods can be seen as a variation (or a particular case) of cost-sensitive learning algorithms largely presented in machine learning where one associates a learning cost to each instance and uses it to minimize a new weighted loss function. An important thing that we should note here is that this procedure produces an optimal model for the target domain only in the case when the support of $\mathcal{T}_{\mathbf{X}}$ is contained in the support of $\mathcal{S}_{\mathbf{X}}$. Even though this assumption may seem too restrictive and hard to meet in real-world applications, we further present a result showing that, even in this case, a covariate shift problem can be really hard to solve.

2.2.2. *Target shift*

Second scenario, which we consider below, is called *target shift*. This particular scenario deals with situations occurring in $Y \rightarrow \mathbf{X}$ problems where the shift is due to change in the target distributions of source and target domains, i.e. $\mathcal{S}_Y \neq \mathcal{T}_Y$. This form of transfer learning arises frequently in applications where training and testing data are gathered according to different sampling strategies. For example, training data can be collected using an established plan that respects equal (or any desired) proportions of classes of interest while the test data were collected *a posteriori* without any control. We illustrate this in Figure 2.4.

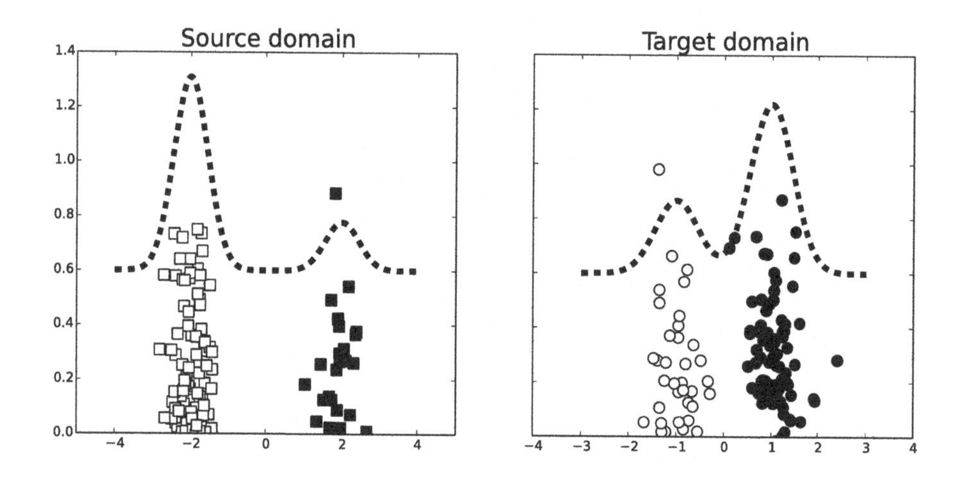

Figure 2.4. *Illustration of target shift. Here, marginal distributions of outputs in source domains (left) and target domains (right) are different while the conditional distributions remain the same. This problem can be seen as a learning problem with class imbalance*

In real-world applications, the target shift often occurs in regression problems where the output variable exhibits a significant change in the magnitude of its values. For instance, in [ZHA 13], the real-world data set used to model the target shift problem was given by a non-stationary time series related to a cause–effect pair with $\mathbf{X}$ being the number of bytes sent by a computer at the ith minute while Y being the number of open http connections at the same time. In this case, the target shift problem can be obtained by dividing the data into two subsets where the source and target domains are represented by intervals with high and low numbers of open http connections, respectively. This is illustrated in Figure 2.5.

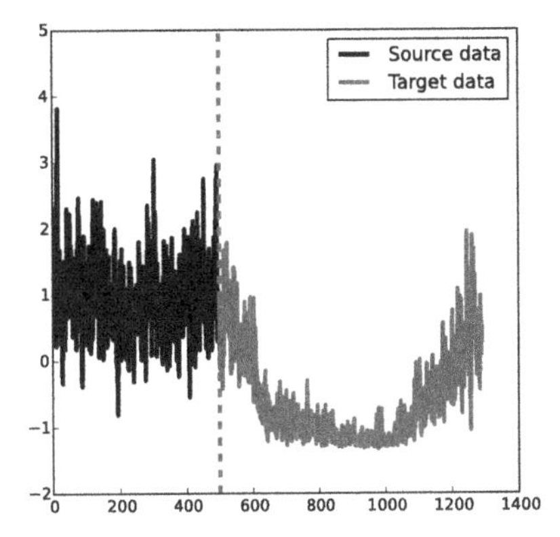

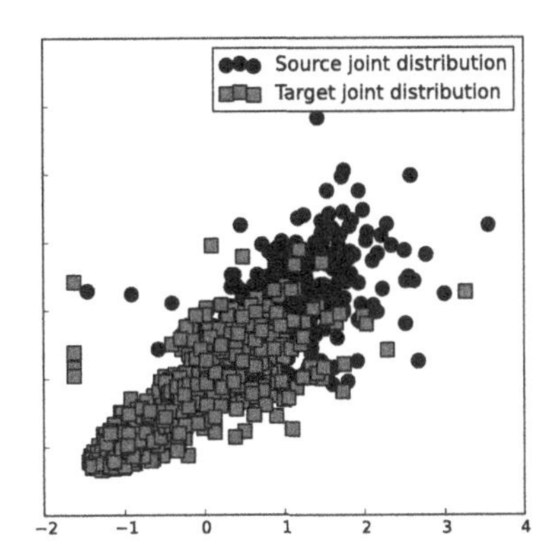

Figure 2.5. *Non-stationary time series (left) and the joint source and target distributions (right) illustrating the target shift problem. Here, the variables are supposed to have a direct causal relation, so that the conditional distributions of both variables are likely to be the same*

To formalize the target shift problem, without the loss of generality, let us consider a binary classification problem with a source sample being drawn from a probability distribution

$$\mathcal{S}_{\mathbf{X}} := (1 - \pi_S)\mathcal{P}_0 + \pi_S\mathcal{P}_1,$$

where $0 < \pi_S < 1$ and $\mathcal{P}_0, \mathcal{P}_1$ are marginal distributions of the source data given the class labels 0 and 1, respectively, with $\mathcal{P}_0 \neq \mathcal{P}_1$. The marginal distribution of the target domains can also be defined as follows

$$\mathcal{T}_{\mathbf{X}} = (1 - \pi_T)\mathcal{P}_0 + \pi_T\mathcal{P}_1,$$

so that $\exists\, j \in [1, \ldots, N] : \pi_S^j \neq \pi_T$. This last condition is a characterization of target shift used in previous theoretical works on the subject [BLA 10, SAN 14, SCO 13].

The key result that allows us to solve the target shift problem was given by Blanchard *et al.* [BLA 10] and is related to the irreducibility of one distribution with respect to another. Its definition is given below.

DEFINITION 2.2.– *Let $\mathcal{G}$, $\mathcal{H}$ be probability distributions. We say that $\mathcal{G}$ is irreducible with respect to $\mathcal{H}$ if there exists no decomposition of the form $\mathcal{G} = \gamma\mathcal{H} + (1-\gamma)\mathcal{F}_0$, where $\mathcal{F}_0$ is some probability distribution and $0 < \gamma \leq 1$. We say that $\mathcal{G}$ and $\mathcal{H}$ are mutually irreducible if $\mathcal{G}$ is irreducible with respect to $\mathcal{H}$ and vice versa.*

The following result was proved by Blanchard *et al.* [BLA 10] for this case.

PROPOSITION 2.1.– *Let $\mathcal{F}$, $\mathcal{H}$ be probability distributions. If $\mathcal{F} \neq \mathcal{H}$, there is a unique $\nu^* \in [0;1)$ and G such that $\mathcal{F} = (1-\nu^*)\mathcal{G} + \nu^*\mathcal{H}$, and such that $\mathcal{G}$ is irreducible with respect to $\mathcal{H}$. If we additionally define $\nu^* = 1$ when $\mathcal{F} = \mathcal{H}$, then in all cases $\nu^*(\mathcal{F},\mathcal{G}) = \max\{\alpha \in [0;1] : \exists\, \mathcal{G}'$probability distribution $\mathcal{F} = (1-\alpha)\mathcal{G}' + \alpha\mathcal{H}\}$.*

This proposition applied in the domain adaptation context defined above allows us to define the unknown class proportions π_S and π_T based on $\nu^*(\mathcal{F},\mathcal{G})$. Once this is done, one can easily use a reweighting technique explained above for the case of covariate shift to allow efficient learning. This solution, however, holds only in cases when every element of conv$\{\mathcal{P}_l : l \notin \{0,1\}\}$ is irreducible with respect to $\mathcal{P}_0$ and $\mathcal{P}_1$ where conv denotes the set of convex combinations distributions, that is, the set of mixture distributions based on them.

Also, note that the connection between the instance-reweighting algorithm and the solution of the target shift problem presented above exhibit another important similarity. Indeed, while in the covariate-shift scenario the optimal estimate of weights is equal to the estimate of $\frac{\mathcal{T}_{\mathbf{X}}(\mathbf{x})}{\mathcal{S}_{\mathbf{X}}(\mathbf{x})}$. For target shift Scott *et al.* [SCO 13] showed that $\nu^*(\mathcal{F},\mathcal{G})$ is equal to the essential infimum of $\frac{f(\mathbf{x})}{h(\mathbf{x})}$ where f and h are densities of $\mathcal{F}$ and $\mathcal{G}$, respectively.

Finally, a somehow different analysis of target shift was proposed in [ZHA 13]. The authors considered the learning problem under the following set of assumptions:

1) $\text{SUPP}(\mathcal{T}_Y) \subseteq \text{SUPP}(\mathcal{S}_Y)$ meaning that the source domain is richer than the target domain;

2) there exists only one possible distribution of Y that, together with $\mathcal{S}(\mathbf{x}|y)$, leads to $\mathcal{T}_{\mathbf{X}}$.

Under these assumptions, the algorithm proposed by the authors based on the minimization of the discrepancy between the kernel embeddings of the source and target marginal distributions leads to a solution that provably converges to $\frac{\mathcal{T}_Y}{\mathcal{S}_Y}$. Once the solution is obtained, it can be used directly in the instance-reweighting scheme as explained above.

2.2.3. *Concept shift*

Above, we considered two different adaptation scenarios related to shifting marginal distributions of instances and outputs while assuming that the conditional distributions remained equal across domains. We now proceed by introducing the *concept shift*, an adaptation problem that occurs in both $\mathbf{X} \to Y$ and $Y \to \mathbf{X}$ problems characterized by the shift in the conditional distributions $\mathcal{S}(Y|\mathbf{X}) \neq \mathcal{T}(Y|\mathbf{X})$ and $\mathcal{S}(\mathbf{X}|Y) \neq \mathcal{T}(\mathbf{X}|Y)$, respectively. A graphical representation of concept shift for a $Y \to \mathbf{X}$ problem is given in Figure 2.6.

This particular adaptation scenario is considered as one of the most difficult ones which explains a limited amount of work that tackles the above-mentioned problem. Among the approaches that were proposed to deal with it, we can highlight the conditional matching algorithm and a two-stage offset algorithms proposed in [WAN 14] for the $\mathbf{X} \rightarrow Y$ problem and the conditional matching algorithm proposed in [ZHA 13] for the $Y \rightarrow \mathbf{X}$ problem. Both of these approaches are based on the mean kernel embeddings of conditional probability distributions where the correction is performed over the transformed outputs and instance spaces, respectively. As both methods consider the same affine transformation, we only present the contribution of the original paper [ZHA 13] below.

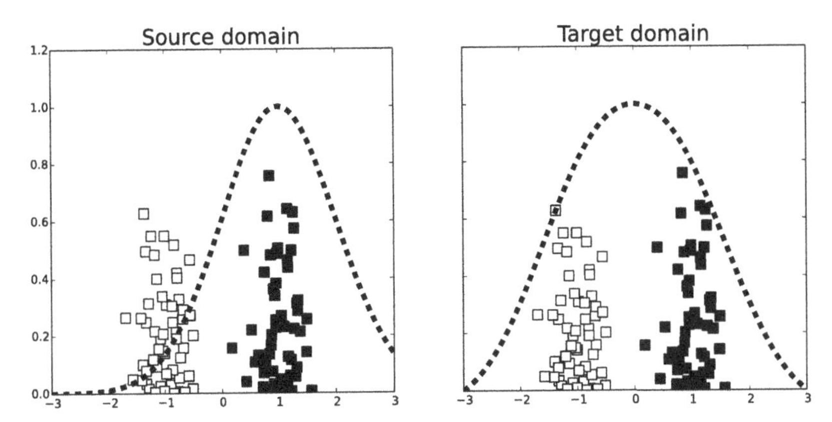

Figure 2.6. *Illustration of concept shift. Here, marginal distributions in source domains (left) and target domains (right) are the same while the conditional distributions are different*

We propose to define a location-scale transformation of the source sample S_u as follows:

$$S^{\text{new}} = S_u \odot W + B.$$

Based on this new sample, we can further define a transformed source distribution $\mathcal{S}_{\mathbf{X}}^{\text{new}}$ and then aim to minimize the following quantity over the space of matrices W and B:

$$\min_{W,B} ||\mu[\mathcal{S}_{\mathbf{X}}^{\text{new}}] - \mu[\mathcal{T}_{\mathbf{X}}]||,$$

where $\mu[\cdot]$ stands for a kernel embedding of a probability distribution that we define in the following chapter. This algorithm was proved to lead to a perfect matching between conditional distributions for both $\mathbf{X} \rightarrow Y$ and $Y \rightarrow \mathbf{X}$ problems under the assumption that the affine transformation between two conditional distributions exists. It is important to note that, contrary to two previous cases considered above, the instance reweighting approach is replaced by a procedure where one learns a classifier directly on the transformed source samples and applies it in the target domain. This type of approach is related to feature-transformation approaches, which are widely studied nowadays due to the success of various representation learning algorithms, including deep learning methods.

2.2.4. *Sample-selection bias*

Sample-selection bias is a problem often encountered in various applications where the source and target distributions differ due to the presence of a latent variable that tends to exclude some observations from the sample depending on their class label or their nature. In the first case, one can think of a classification task with several classes of interest representing handwritten digits. For this task, a simple pre-processing consisting of eliminating instances written unclearly can lead to a sample-selection bias as, in general, some digits are more likely to be written unclearly than others. Regarding the second case, we can think of a street survey that takes place on a weekday before noon. Obviously, this survey will tend to exclude people that work as they will most likely be in their office at that time. In both cases, however, the results of the survey will be considered as general and applied to all groups of people and all classes of interest, thus excluding the effect of the latent variable.

From the examples presented above, we can formalize the sample-selection bias problem as follows. Given a random binary variable ξ that depends on both $\mathbf{X}, Y$, the source and target distributions can be written as follows:

$$\mathcal{S}(\mathbf{X}, Y) = \mathbf{Pr}\,[\xi = 1|\mathbf{X}, Y]\mathbf{Pr}\,[\mathbf{X}]\mathbf{Pr}\,[Y|\mathbf{X}],$$

$$\mathcal{T}(\mathbf{X}, Y) = \mathbf{Pr}\,[\mathbf{X}]\mathbf{Pr}\,[Y|\mathbf{X}],$$

where variable ξ plays the role of the selection function that tends to include or exclude observations based on their label or covariate description. Note also that when ξ does not depend on Y, this problem reduces to the covariate shift presented above, as in this case the joint source distribution can be written as:

$$\mathcal{S}(\mathbf{X}, Y) = \mathbf{Pr}\,[\xi = 1|\mathbf{X}]\mathbf{Pr}\,[\mathbf{X}]\mathbf{Pr}\,[Y|\mathbf{X}] = \tilde{\mathbf{Pr}}[\mathbf{X}]\mathbf{Pr}\,[Y|\mathbf{X}],$$

leading to the inequality between marginal distributions

$$\mathcal{T}_{\mathbf{X}} = \mathbf{Pr}\,[\mathbf{X}] \neq \tilde{\mathbf{Pr}}[\mathbf{X}] = \mathcal{S}_{\mathbf{X}}.$$

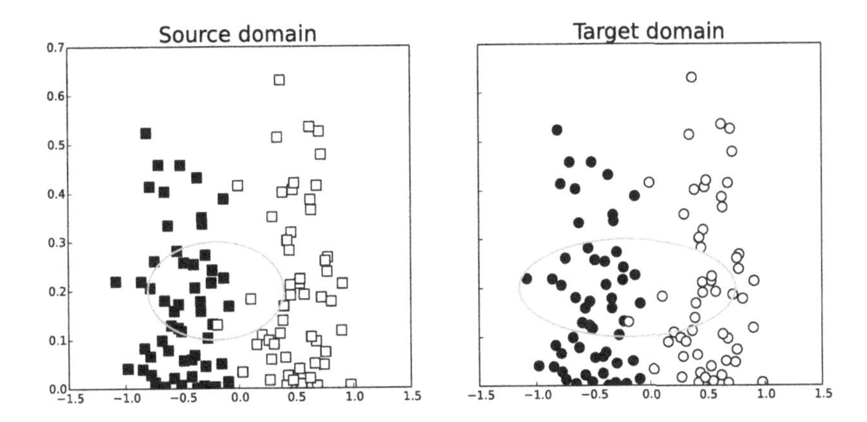

Figure 2.7. *Illustration of sample selection bias. Here, gray ellipses show the decision function of the selection variable ξ. Note that in sample selection bias, the decision zone depends on both the class label and the data point itself. For a color version of this figure, see www.iste.co.uk/redko/domain.zip*

2.2.5. *Other adaptation scenarios*

While the presented adaptation problems are among the most frequently studied ones, we note, however, that some surveys also mention other domain adaptation problems that are not covered in this chapter. These generally include the following:

1) Domain shift: a situation characterized by the change in the measurement system or in the method of description. This particular scenario can be formalized as follows: consider an underlying unchanging latent representation of the covariate space $\mathbf{X}_0$ that is mapped by a certain function f to provide the observed samples from $\mathbf{X}$. In this case, the domain shift is a situation when the mapping f changes between two domains. This change in mapping essentially leads to the change in the marginal distributions of the observed samples from source and target domains even though the real change happens at a deeper level. From this observation, it becomes clear that this problem can be solved using the same techniques as the covariate shift problem even though the two are fundamentally different from the theoretical point of view.

2) Source component shift: an adaptation scenario where the observed data are assumed to be composed of a certain number of different components with the proportions of components that vary between the source and target domains. This situation can be further characterized based on the origin of the composition structure: when the source and target distributions are mixtures with different proportions, we talk about the *mixture component shift*; when each covariate is a mixture of covariates drawn from different sources, we deal with *mixing component shift*; finally, when data are dependent on a number of factors that can be decomposed into a form and

a strength, we consider the *factor component shift*. As in source component shift, there exist an explicit dependence of data on some unknown latent variable. This particular adaptation problem can thus be generally converted to sample-selection bias and considered using the methods suitable for it.

2.3. Necessity of a divergence between domains

In domain adaptation, we often talk about the similarity between source and target domains, but so far we have not defined it properly. Obviously, this notion of similarity should play a crucial role in the theoretical analysis of domain adaptation, as the discrepancy between the training and test data differentiate transfer learning from traditional supervised learning and is therefore of great importance. In order to give an indication of how this similarity can be defined, let us consider an example from multilingual adaptation where the objective is to classify text documents in different languages based on their content. In this particular situation, the most intuitive way to define similarity is to assess the proximity of languages by taking into account the proximity of the vocabularies used in each set of documents. This generally involves comparing the presence of words carrying similar information and their frequency of occurrence. The above-mentioned comparison leads us to the idea of representing a document as a set of frequencies of different words: a construction that is very similar to the discrete probability distribution. In this case, the measure of similarity can be considered as a divergence between the probability distributions of two domains with each of them being represented by a document in a different language. This example is used in the context of multilingual adaptation, but it applies to many different tasks as well. For instance, in visual domain adaptation, it is sufficient to replace the notion of frequency of a word in a document with the intensity of a pixel in an image and the argument presented above would still hold. Finally, we note that in a classification task, each document or image has its own associated label. In this case, the notion of probability distribution extends to the joint probability distribution on the product space of the considered instances and the output space of their labels.

The idea presented above spurred the divergence-based generalization bounds in a domain adaptation context: probably the most extensive family of results in domain adaptation that builds upon the idea of bounding the gap between the source and target domains' error functions in terms of the discrepancy between their probability distributions. As explained previously, different assumptions can be made regarding the nature of the learning problem, resulting in different adaptation scenarios. Basically, the main assumption done in the following results is that the two domains are "sufficiently" related according to their labels: if the true labeling function of the source domain is too different from the true labeling function of the target domain, then one cannot expect it to adapt well. The elementary building block usually used to derive divergence-based generalization bounds can often be expressed as the following result:

$$\left|\mathrm{R}^{\ell}_{\mathcal{T}}(h,h') - \mathrm{R}^{\ell}_{\mathcal{S}}(h,h')\right| \leq \mathrm{dist}(\mathcal{S}_{\mathbf{X}}, \mathcal{T}_{\mathbf{X}}),$$

where $(h, h') \in \mathcal{H}^2$ are two hypotheses and $\mathrm{dist}(\cdot,\cdot)$ is a divergence measure defined on the space of probability distributions, and

$$\mathrm{R}^{\ell}_{\mathcal{D}}(h,h') = \mathop{\mathbf{E}}_{\mathbf{x}\sim\mathcal{D}_{\mathbf{X}}} \ell(h(\mathbf{x}), h'(\mathbf{x})).$$

Bearing in mind this general form, one can ask two important questions that will be in the core of the results presented below. These questions are:

– what divergences from the abundant amount of the existing metrics defined on the space of probability measures can be used in domain adaptation?

– can we design new divergence measures suitable for domain adaptation?

Answering the first question is generally equivalent to proving that a certain divergence measure can be used to relate the errors in the source and target domains. Indeed, if this is the case, one can often directly derive an adaptation algorithm based on the minimization of the considered divergence between the observed samples that will probably lead to efficient adaptation. Furthermore, introducing existing divergence measures allows us to compare them in order to analyze the tightness of the bounds and thus to access their relative efficiency in domain adaptation.

Another way to proceed, suggested by the second question, is to introduce new divergence measures suitable for domain adaptation. This can be done, for instance, by modifying existing divergence measures in order to link them to a classification or regression problem at hand or by introducing principally new divergence measures that capture the varying nature of the considered domains in a certain sense. The first approach can be used when a given divergence measure is known to be too loose and/or hard to estimate in practice, which means that a certain customization is needed in order to make it useful. This is the case, for instance, for a popular L^1-distance (Definition 3.1 below) that can take vacuous values in practice leading to uninformative results. On the other hand, the second approach can be used when one pays a particular attention to a certain aspect that causes the shift between the two domains and wants to quantify it using a special discrepancy measure. In both cases, however, introducing new metrics for domain adaptation requires an additional analysis addressing a consistent strategy for its finite-sample approximation, which may be quite a challenging problem in its own right.

2.4. Summary

In this chapter, we gave a brief overview of the main domain adaptation scenarios that can be found in the literature. For each of the scenarios considered, we gave an algorithmic idea that can be used to solve the underlying problem and several hints that justify the appropriateness of the solution from the theoretical point of view. In the following chapters, we recall the main theoretical analyses done so far on domain adaptation in order to give a broad overview of different domain adaptation generalization guarantees.

For the general target shift problem, the generalization guarantees can be derived from the work of [BLA 10] when expressing the target shift problem as a multiclass novelty detection problem, as it is done in [SAN 14]. In what follows, we omit these results and limit ourselves only to a brief account concerning them provided in section 2.2.2. Regarding the sample-selection bias problem, we note that it is usually reduced to either the target shift problem or the covariate shift problem when the latent variable that decides whether an instance will or will not be included in the training sample only depends on the covariates or on labels. To the best of our knowledge, there are no results for the sample-selection bias problem in the literature that consider the mutual dependence on both variables at the same time. As for the concept shift, this problem is usually considered to be extremely hard to solve in practice, leading to a very low number of works on the subject: its theoretical analysis is roughly limited to two papers cited in section 2.2.3.

3

Seminal Divergence-based Generalization Bounds

In this chapter, we give a comprehensive description of different generalization bounds for domain adaptation that are based on divergence measures between source and target probability distributions. Before we proceed to the formal presentation of the mathematical results, we first explain why this particular type of result presents the vast majority of all available domain adaptation results. We further take a closer look at the very first steps that were established in order to provide a theoretical background for domain adaptation. Surprisingly, these first results reflected in a very direct way the general intuition behind domain adaptation and remained quite lose to the traditional generalization inequalities presented in Chapter 1. After this, we turn our attention to different strategies that were proposed in order to overcome the flaws of the seminal results as well as to strengthen the obtained theoretical guarantees.

3.1. The first theoretical analysis of domain adaptation

From a theoretical point of view, the domain adaptation problem was investigated for the first time by Ben-David *et al.* [BEN 07, BEN 10a][1]. The authors of these papers focused on the domain adaptation problem following the uniform convergence theory (recalled in section 1.2.1) and considered the $0-1$ loss (equation [1.1]) function in the setting of binary classification with $Y = \{-1, +1\}$. In this work, they proposed to make use of the L^1-*distance*, the definition of which is given below.

1 Note that in [BEN 10a] the authors presented an extended version of the results previously published in [BEN 07] and [BLI 08].

DEFINITION 3.1 (L^1-distance).– *Denote by $\mathcal{B}$ the set of measurable subsets under two probability distributions $\mathcal{D}_1$ and $\mathcal{D}_2$. Then the L^1-distance or the total variation distance between $\mathcal{D}_1$ and $\mathcal{D}_2$ is defined as*

$$d_1(\mathcal{D}_1, \mathcal{D}_2) = 2 \sup_{B \in \mathcal{B}} \left| \Pr_{\mathcal{D}_1}(B) - \Pr_{\mathcal{D}_2}(B) \right|.$$

The L^1-distance, also known as total variation distance, is a proper metric on the space of probability distributions that informally quantifies the largest possible difference between the probabilities that the two probability distributions $\mathcal{D}_1$ and $\mathcal{D}_2$ can assign to the same event B. This distance is quite popular in many real-world applications such as image denoising or numerical approximations of partial differential equations.

3.2. One seminal result

Starting from definition 3.1, the first important result proved in their work was formulated as follows.

THEOREM 3.1 ([BEN 07]).– *Given two domains $\mathcal{S}$ and $\mathcal{T}$ over $\mathbf{X} \times Y$ and a hypothesis class $\mathcal{H}$, the following holds for any $h \in \mathcal{H}$*

$$\forall h \in \mathcal{H}, \quad \mathrm{R}_{\mathcal{T}}^{\ell_{01}}(h) \leq \mathrm{R}_{\mathcal{S}}^{\ell_{01}}(h) + d_1(\mathcal{S}_{\mathbf{X}}, \mathcal{T}_{\mathbf{X}})$$
$$+ \min \left\{ \mathop{\mathbf{E}}_{\mathbf{x} \sim \mathcal{S}_{\mathbf{X}}} \left[|f_{\mathcal{S}}(\mathbf{x}) - f_{\mathcal{T}}(\mathbf{x})| \right], \mathop{\mathbf{E}}_{\mathbf{x} \sim \mathcal{T}_{\mathbf{X}}} \left[|f_{\mathcal{T}}(\mathbf{x}) - f_{\mathcal{S}}(\mathbf{x})| \right] \right\},$$

where $f_{\mathcal{S}}(\mathbf{x})$ and $f_{\mathcal{T}}(\mathbf{x})$ are source and target true labeling functions associated with $\mathcal{S}$ and $\mathcal{T}$, respectively.

The proof of this theorem can be found in Appendix 1. This theorem presents a first result that relates the performance of a given hypothesis function with respect to two different domains. It implies that the error achieved by a hypothesis in the source domain upper bounds the true error on the target domain where the tightness of the bound depends on the distance between their distributions and that of labeling functions. Despite being the first result of this kind proposed in the literature, the idea of bounding the error in terms of the L^1-distance between marginal distributions of the two domains presents two important restrictions:

1) the L^1-distance cannot be estimated from finite samples for arbitrary probability distributions;

2) it does not allow us to link the divergence measure to the considered hypothesis class and thus leads to very loose inequality.

In order to address these issues, the authors further defined the $\mathcal{H}\Delta\mathcal{H}$-divergence based on the $\mathcal{A}$-divergence introduced by Kifer *et al.* [KIF 04] for data stream change detection. We give its definition below.

DEFINITION 3.2 (Based on [KIF 04]).– *Given two domains' marginal distributions $\mathcal{S}_{\mathbf{X}}$ and $\mathcal{T}_{\mathbf{X}}$ over the input space $\mathbf{X}$, let $\mathcal{H}$ be a hypothesis class, and denote $\mathcal{H}\Delta\mathcal{H}$ the symmetric difference hypothesis space defined as*

$$h \in \mathcal{H}\Delta\mathcal{H} \iff h(\mathbf{x}) = g(\mathbf{x}) \oplus g'(\mathbf{x})$$

for some $(g, g')^2 \in \mathcal{H}^2$, where $\oplus$ stands for XOR operation. Let $I(h)$ denote the set for which $h \in \mathcal{H}\Delta\mathcal{H}$ is the characteristic function, that is $\mathbf{x} \in I(h) \Leftrightarrow g(\mathbf{x}) = 1$. The $\mathcal{H}\Delta\mathcal{H}$-divergence between $\mathcal{S}_{\mathbf{X}}$ and $\mathcal{T}_{\mathbf{X}}$ is defined as:

$$d_{\mathcal{H}\Delta\mathcal{H}}(\mathcal{S}_{\mathbf{X}}, \mathcal{T}_{\mathbf{X}}) = 2 \sup_{h \in \mathcal{H}\Delta\mathcal{H}} \left| \mathop{\mathbf{Pr}}_{\mathcal{S}_{\mathbf{X}}}(I(h)) - \mathop{\mathbf{Pr}}_{\mathcal{T}_{\mathbf{X}}}(I(h)) \right|.$$

The $\mathcal{H}\Delta\mathcal{H}$-divergence solves both problems associated with the L^1-distance. First, from its definition we can instantly see that $\mathcal{H}\Delta\mathcal{H}$-divergence takes into account the considered hypothesis class. This association ensures that the bound remains meaningful and directly related to the learning problem at hand. On the other hand, the $\mathcal{H}\Delta\mathcal{H}$-divergence for any class $\mathcal{H}$ is never larger than the L^1-distance; thus it can lead to a tighter inequality in the bound. Furthermore, for a given hypothesis class $\mathcal{H}$ of finite VC dimension, the $\mathcal{H}\Delta\mathcal{H}$-divergence can be estimated from finite samples using the following lemma.

LEMMA 3.1.– *Let $\mathcal{H}$ be a hypothesis space of VC dimension $VC(\mathcal{H})$. If S_u, T_u are unlabeled samples of size m each, drawn independently from $\mathcal{S}_{\mathbf{X}}$ and $\mathcal{T}_{\mathbf{X}}$ respectively, then for any $\delta \in (0, 1)$ with probability at least $1 - \delta$ over the random choice of the samples we have*

$$d_{\mathcal{H}\Delta\mathcal{H}}(\mathcal{S}_{\mathbf{X}}, \mathcal{T}_{\mathbf{X}}) \leq \hat{d}_{\mathcal{H}\Delta\mathcal{H}}(S_u, T_u) + 4\sqrt{\frac{2\, VC(\mathcal{H}) \log(2m) + \log(\frac{2}{\delta})}{m}}, \qquad [3.1]$$

where $\hat{d}_{\mathcal{H}\Delta\mathcal{H}}(S_u, T_u)$ is the empirical $\mathcal{H}\Delta\mathcal{H}$-divergence estimated on S_u and T_u.

Inequality [3.1] shows that with an increasing number of instances and for a hypothesis class of a finite VC dimension, the empirical $\mathcal{H}\Delta\mathcal{H}$-divergence can be a good proxy for its true counterpart. The former can be further calculated because of the following result.

LEMMA 3.2 ([BEN 10a]).– *Let $\mathcal{H}$ be a hypothesis space. Then, for two unlabeled samples S_u, T_u of size m we have*

$$\hat{d}_{\mathcal{H}\Delta\mathcal{H}}(S_u, T_u) = 2\left(1 - \min_{h \in \mathcal{H}\Delta\mathcal{H}}\left[\frac{1}{m}\sum_{\mathbf{x}:h(\mathbf{x})=0}\mathbf{I}\left[\mathbf{x} \in S_u\right] + \frac{1}{m}\sum_{\mathbf{x}:h(\mathbf{x})=1}\mathbf{I}\left[\mathbf{x} \in T_u\right]\right]\right).$$

The proof of this lemma can be found in Appendix 1.

One may note that the expression of the empirical $\mathcal{H}\Delta\mathcal{H}$-divergence given above is essentially the error of the best classifier for the binary classification problem of distinguishing between the source and target instances pseudo-labeled with 0's and 1's. In practice, it means that the value of the $\mathcal{H}\Delta\mathcal{H}$-divergence depends explicitly on the hypothesis class used to produce a classifier. This dependence, as well as the intuition behind the $\mathcal{H}\Delta\mathcal{H}$-divergence, is illustrated in Figure 3.1.

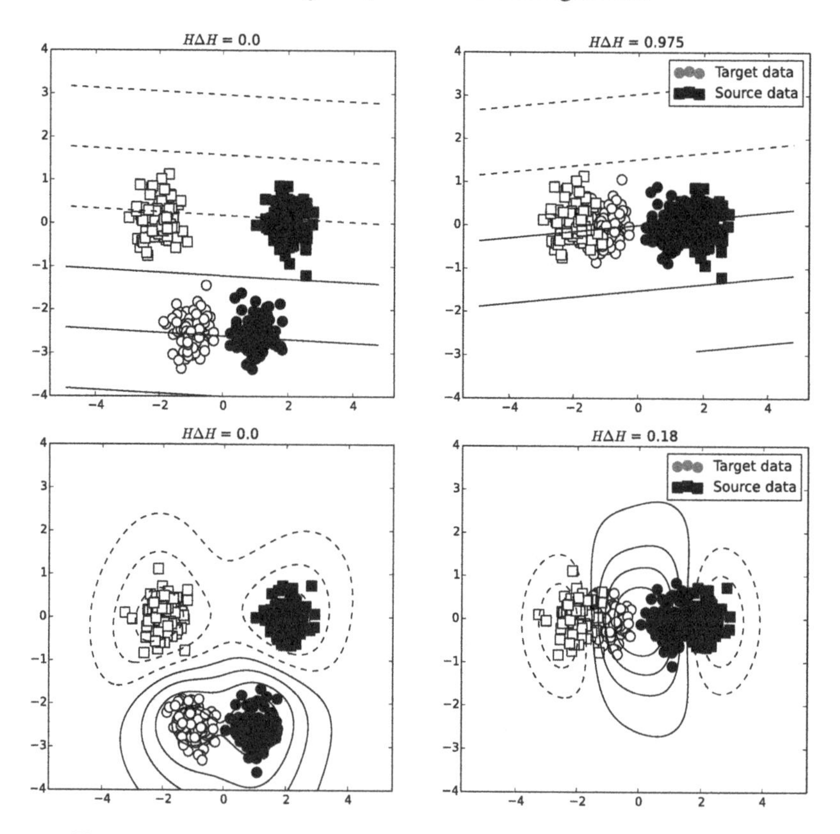

Figure 3.1. *Illustration of the $\mathcal{H}\Delta\mathcal{H}$-divergence when the hypothesis class consists of linear (top row) and nonlinear (bottom row) classifiers*

In this figure, we consider two different domain adaptation problems where for one of them the source and target samples are well separated, while for the other the source and target data are mixed together. In order to calculate the value of the $\mathcal{H}\Delta\mathcal{H}$-divergence, we need to choose a hypothesis class used to produce a classifier. Here, we consider two different families of classifiers where the first one corresponds to a linear support vector machines (SVM) classifier while the second one is given by its nonlinear version with radial basis function (RBF) kernels. For each solution, we also plot the decision boundaries in order to see how the source and target instances are classified in both cases. From the visualization of the decision boundaries, we note that the linear classifier fails to distinguish between the mixed source and target instances, while the nonlinear classifier manages to do it quite well. This is reflected in the value of the $\mathcal{H}\Delta\mathcal{H}$-divergence which is equal to zero in the first case for both classifiers and is drastically different for the second adaptation problem. Having two different divergence values for the same adaptation problem may seem surprising at the first sight but it has a simple explanation. By choosing a richer hypothesis class composed of nonlinear functions, we increased the VC dimension of the considered hypothesis space and thus increased the complexity term in Lemma 3.1. This shows the trade-off that one has to bear in mind when calculating the $\mathcal{H}\Delta\mathcal{H}$-divergence in the same way as is suggested by the general VC theory.

3.3. A generalization bound for domain adaptation

At this point, we already have a "reasonable" version of the L^1-distance used to derive the first seminal result and we presented its finite sample approximation, but we have not yet applied it to relate the source and target error functions. The following lemma gives the final key needed to obtain a generalization bound for domain adaptation that is linked to a specific hypothesis class and is derived for available source and target finite size samples. It reads as follows.

LEMMA 3.3 ([BEN 10a]).– *Let $\mathcal{S}$ and $\mathcal{T}$ be two domains on $\mathbf{X} \times Y$. For any pair of hypotheses $(h, h') \in \mathcal{H}\Delta\mathcal{H}^2$, we have*

$$\left| \mathrm{R}_{\mathcal{T}}^{\ell_{01}}(h, h') - \mathrm{R}_{\mathcal{S}}^{\ell_{01}}(h, h') \right| \leq \frac{1}{2} d_{\mathcal{H}\Delta\mathcal{H}}(\mathcal{S}_{\mathbf{X}}, \mathcal{T}_{\mathbf{X}}).$$

Note that in this lemma, for which the proof can be found in Appendix 1, the source and target risk functions are defined for the same pairs of hypotheses while the true risk should be calculated based on a given hypothesis and the corresponding labeling function. This result, presenting the complete generalization bound for domain adaptation with $\mathcal{H}\Delta\mathcal{H}$-divergence, is established by the means of the following theorem.

THEOREM 3.2 ([BEN 10a]).– *Let $\mathcal{H}$ be a hypothesis space of VC dimension $VC(\mathcal{H})$. If S_u, T_u are unlabeled samples of size m' each, drawn independently from $\mathcal{S}_{\mathbf{X}}$ and*

$\mathcal{T}_\mathbf{X}$, respectively, then for any $\delta \in (0,1)$ with probability at least $1-\delta$ over the random choice of the samples, we have that for all $h \in \mathcal{H}$

$$\mathrm{R}_{\mathcal{T}}^{\ell_{01}}(h) \leq \mathrm{R}_{\mathcal{S}}^{\ell_{01}}(h) + \tfrac{1}{2}\hat{d}_{\mathcal{H}\Delta\mathcal{H}}(S_u, T_u) + 4\sqrt{\frac{2\,VC(\mathcal{H})\log(2m') + \log(\frac{2}{\delta})}{m'}} + \lambda\,,$$

where λ is the combined error of the ideal hypothesis h^* that minimizes $\mathrm{R}_{\mathcal{S}}(h) + \mathrm{R}_{\mathcal{T}}(h)$.

The proof of this theorem can be found in Appendix 1. As pointed out at the beginning of this chapter, a meaningful domain adaptation generalization bound should include two terms that reflect both the divergence between the marginal distribution of the source and target domains and the divergence between the labeling functions. The first term here is obviously reflected by the $\mathcal{H}\Delta\mathcal{H}$-divergence between the observable samples, while the second one is given by the λ term since it depends on the true labels (and can be seen as a measure of capacity to adapt). It is important to note that without information on the target labels it is impossible to estimate λ, implying that the bound can be very loose.

The presence of the trade-off between source risk, divergence and capability to adapt, is a very important phenomenon in domain adaptation. Indeed, it shows that reducing the divergence between the samples can be insufficient when there is no hypothesis that can achieve a low error on both source and target samples. In other words, a good hypothesis for the source domain could be a good hypothesis for the target domain if the two domains are very close. This spurred one of the main ideas used to derive novel domain adaptation algorithms: assuming that λ is low (i.e. a good classifier for both domains exists), we need to find a good representation space where the two domains are close, while maintaining good performance in the source domain.

3.4. Generalization bounds for the combined error

In the unsupervised case that we have considered previously, it is assumed that we have no access to labeled instances in the target domain that can help to guide adaptation. For this case, the main strategy leading to efficient adaptation is to learn a classifier on a target-aligned labeled sample from the source domain and to apply it directly in the target domain afterwards. While this situation occurs quite often in practice, many examples can be found where several labeled target instances are available during the learning stage. In what follows, we consider this situation and give a generalization bound for it, showing that the error obtained by a classifier learned from a mixture of source and target labeled data can be upper bounded by the error of the best classifier learned using the target domain data only.

To proceed, let us now assume that we possess βm instances drawn independently from $\mathcal{T}$ and $(1-\beta)m$ instances drawn independently from $\mathcal{S}$ and labeled by $f_{\mathcal{S}}$ and $f_{\mathcal{T}}$, respectively. A natural goal for this setting is to use the available labeled instances from the target domain to find a trade-off between minimizing the source and the target errors depending on the number of instances available in each domain and the distance between them. In this case, we can consider the empirical combined error [BLI 08] defined as a convex combination of errors in the source and target training data:

$$\hat{\mathrm{R}}^{\alpha}(h) = \alpha \mathrm{R}^{\ell_{01}}_{\hat{\mathcal{T}}}(h) + (1-\alpha)\mathrm{R}^{\ell_{01}}_{\hat{\mathcal{S}}}(h),$$

where $\alpha \in [0,1]$.

The use of the combined error is motivated by the fact that if the number of instances in the target sample is small compared to the number of instances in the source domain (which is usually the case in domain adaptation), minimizing only the target error may not be appropriate. Instead, we may want to find a suitable value of α that ensures the minimum of $\mathrm{R}^{\alpha}(h)$ with respect to a given hypothesis h. Note that in this case the shape of the generalization bound that we are interested in becomes different. Indeed, in all previous theorems the goal is to upper bound the target error by the source error, while in this case we would like to know if learning a classifier by minimizing the combined error is better than minimizing the target error using the available labeled instances alone. The answer to this question is given by the following theorem.

THEOREM 3.3 ([BLI 08, BEN 10a]).– *Let $\mathcal{H}$ be a hypothesis space of VC dimension $VC(\mathcal{H})$. Let $\mathcal{S}$ and $\mathcal{T}$ be the source and target domain, respectively, defined on $\mathbf{X}\times Y$. Let S_u, T_u be unlabeled samples of size m' each, drawn independently from $\mathcal{S}_{\mathbf{X}}$ and $\mathcal{T}_{\mathbf{X}}$, respectively. Let S be a labeled sample of size m generated by drawing βm points from $\mathcal{T}$ ($\beta \in [0,1]$) and $(1-\beta)\, m$ points from $\mathcal{S}$ and labeling them according to $f_{\mathcal{S}}$ and $f_{\mathcal{T}}$, respectively. If $\hat{h} \in \mathcal{H}$ is the empirical minimizer of $\hat{\mathrm{R}}^{\alpha}(h)$ on S and $h^*_{\mathcal{T}} = \operatorname{argmin}_{h\in\mathcal{H}} \mathrm{R}^{\ell_{01}}_{\mathcal{T}}(h)$, then for any $\delta \in (0,1)$, with probability at least $1-\delta$ over the random choice of the samples, we have*

$$\mathrm{R}^{\ell_{01}}_{\mathcal{T}}(\hat{h}) \;\leq\; \mathrm{R}^{\ell_{01}}_{\mathcal{T}}(h^*_{\mathcal{T}}) + c_1 + c_2,$$

where

$$\begin{aligned} c_1 &= 4\sqrt{\frac{\alpha^2}{\beta} + \frac{(1-\alpha)^2}{1-\beta}}\sqrt{\frac{2\,VC(\mathcal{H})\log(2(m+1)) + 2\log(\frac{8}{\delta})}{m}}, \\ \textit{and } c_2 &= 2(1-\alpha)\left(\frac{1}{2}d_{\mathcal{H}\Delta\mathcal{H}}(S_u,T_u) + 4\sqrt{\frac{2\,VC(\mathcal{H})\log(2m') + \log(\frac{8}{\delta})}{m'}} + \lambda\right). \end{aligned}$$

[3.2]

The proof of this theorem can be found in Appendix 1. This theorem presents an important result that reflects the usefulness of the combined minimization of source and target errors based on the available labeled samples in both domains compared to the minimization of the target error only. It essentially shows that the error achieved by the best hypothesis of the combined error in the target domain is always upper bounded by the error achieved by the best target domain's hypothesis. Furthermore, it implies two important consequences:

1) if $\alpha = 1$, the term related to the $\mathcal{H}\Delta\mathcal{H}$-divergence between the domains disappears as in this case we possess enough labeled data in the target domain and a low-error hypothesis can be produced solely from target data;

2) if $\alpha = 0$, the only way to produce a low-error classifier on the target domain is to find a good hypothesis in the source domain while minimizing the $\mathcal{H}\Delta\mathcal{H}$-divergence between the domains. In this case, we must also assume that λ is low, so that the adaptation is possible.

Additionally, theorem 3.3 can provide some insights about the optimal mixing value of α depending on the quantity of labeled instances in source and target domains. In order to illustrate it, we may rewrite the right-hand side of equation [3.2] as a function of α in order to understand when it is minimized. We write:

$$f(\alpha) = 2B\sqrt{\frac{\alpha^2}{\beta} + \frac{(1-\alpha)^2}{1-\beta}} + 2(1-\alpha)A,$$

where

$$B = \sqrt{\frac{2\,VC(\mathcal{H})\log(2(m+1)) + 2\log(\frac{8}{\delta})}{m}}$$

is a complexity term that approximately equals $\sqrt{VC(\mathcal{H})/m}$, and

$$A = \frac{1}{2}\hat{d}_{\mathcal{H}\Delta\mathcal{H}} S_u T_u + 4\sqrt{\frac{2\,VC(\mathcal{H})\log(2m') + \log(\frac{8}{\delta})}{m'}} + \lambda$$

is the total divergence between two domains.

It then follows that the optimal value α^* is a function of the number of target examples $m_T = \beta m$, the number of source examples $m_S = (1-\beta)m$, and the ratio $D = \sqrt{VC(\mathcal{H})}/A$:

$$\alpha^*(m_S, m_T, D) = \begin{cases} 1, \; m_T \geq D^2 \\ \min(1, \nu), \; m_T \leq D^2 \end{cases}$$

where

$$\nu = \frac{m_T}{m_T + m_S}\left(1 + \frac{m_S}{\sqrt{D^2(m_S + m_T) - m_S m_T}}\right).$$

This reformulation offers a couple of interesting insights. First, if $m_T = 0$ ($\beta = 0$), then $\alpha^* = 0$ and if $m_S = 0$ (i.e. $\beta = 1$), then $\alpha^* = 1$. As mentioned above, it implies that if we have only source or only target data, the most appropriate choice is to use them for learning directly. Second, if the divergence between two domains equals zero, then the optimal combination is to use the training data with uniform weighting of the examples. On the other hand, if there is enough target data, i.e. $m_T \geq D^2 = VC(\mathcal{H})/A^2$, then no source data are required for efficient learning and using, then will hurt the overall performance. This is due to the fact that the possible error decrease brought about by using additional source data is always subject to its increase due to the increasing divergence between the source and target data. Second, for a few target examples, we may not have enough source data to justify its use. In this case, the source domain's sample can simply be ignored. Finally, once we have enough source instances combined with a few target instances, the value for α^* takes intermediate values. This analysis is illustrated by Figure 3.2.

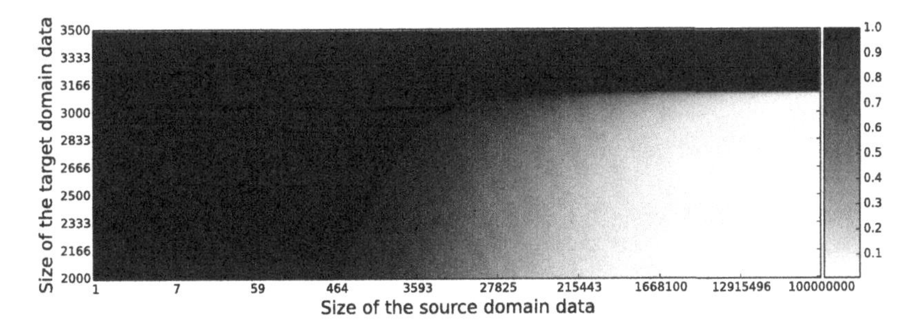

Figure 3.2. *Illustration of the optimal value for α as a function of the size of the source and target domain data*

3.5. Discrepancy distance

One important limitation of the $\mathcal{H}\Delta\mathcal{H}$-divergence is its explicit dependence on a particular choice of loss function, which is taken to be $0-1$ loss. In general, however, we would like to have generalization results for a more general domain adaptation setting where any arbitrary loss function ℓ with some reasonable properties can be considered. In this section, we present a series of results that allow us to extend the first theoretical analysis of domain adaptation presented in the previous section to any arbitrary loss function. As we will show, the new divergence measure considered in this section is not restricted to being used exclusively for the task of binary classification, but it can also be used for large families of regularized classifiers and regression. Moreover, the results of this section use the concept of the Rademacher complexity recalled in Chapter 1. This particular improvement will lead to data-dependent bounds that are usually tighter that the bounds obtained using the VC theory.

3.5.1. *Definition*

We start with the definition of the new divergence measure first introduced by Mansour *et al.* [MAN 09a]. As mentioned by the authors, its name, the *discrepancy distance*, is due to the relationship of this notion with the discrepancy problems arising in combinatorial contexts.

DEFINITION 3.3 ([MAN 09a]).– *Given two domains $\mathcal{S}$ and $\mathcal{T}$ over $\mathbf{X} \times Y$, let $\mathcal{H}$ be a hypothesis class, and let $\ell : Y \times Y \to \mathbb{R}_+$ define a loss function. The discrepancy distance $disc_\ell$ between the two marginals $\mathcal{S}_\mathbf{X}$ and $\mathcal{T}_\mathbf{X}$ over $\mathbf{X}$ is defined by*

$$disc_\ell(\mathcal{S}_\mathbf{X}, \mathcal{T}_\mathbf{X}) = \sup_{(h,h')\in\mathcal{H}^2} \left| \mathop{\mathbf{E}}_{\mathbf{x}\sim\mathcal{S}_\mathbf{X}} \left[\ell\left(h'(\mathbf{x}), h(\mathbf{x})\right)\right] - \mathop{\mathbf{E}}_{\mathbf{x}\sim\mathcal{T}_\mathbf{X}} \left[\ell\left(h'(\mathbf{x}), h(\mathbf{x})\right)\right] \right|.$$

We note that the $\mathcal{H}\Delta\mathcal{H}$-divergence and the discrepancy distance are related. First, for the $0-1$ loss, we have

$$disc_{\ell_{01}}(\mathcal{S}_\mathbf{X}, \mathcal{T}_\mathbf{X}) = \frac{1}{2} d_{\mathcal{H}\Delta\mathcal{H}}(\mathcal{S}_\mathbf{X}, \mathcal{T}_\mathbf{X}),$$

showing that in this case, the discrepancy distance coincides with the $\mathcal{H}\Delta\mathcal{H}$-divergence that appears in theorems 3.2 and 3.3, and suffers from the same computational restrictions as the latter. Furthermore, their tight connection is illustrated by the following proposition.

PROPOSITION 3.1 ([MAN 09a]).– *Given two domains $\mathcal{S}$ and $\mathcal{T}$ over $\mathbf{X} \times Y$, let $\mathcal{H}$ be a hypothesis class and let $\ell : Y \times Y \to \mathbb{R}_+$ define a loss function that is bounded, $\forall (y, y') \in Y^2$, $\ell(y, y') \leq M$ for some $M > 0$. Then, for any hypothesis $h \in \mathcal{H}$, we have*

$$disc_\ell(\mathcal{S}_\mathbf{X}, \mathcal{T}_\mathbf{X}) \leq M\, d_1(\mathcal{S}_\mathbf{X}, \mathcal{T}_\mathbf{X}).$$

The proof of this result can be found in Appendix 1. This proposition establishes a link between the seminal results [BEN 10a] presented in the previous section and shows that for a loss function bounded by M, the discrepancy distance can be upper bounded in terms of the L^1-distance.

3.5.2. *Generalization bounds with discrepancy distance*

In order to present a generalization bound, we first need to understand how the discrepancy distance can be estimated from finite samples. To this end, the authors in [MAN 09a] proposed the following lemma that bounds the discrepancy distance using the Rademacher complexity (see section 1.2.2) of the hypothesis class.

LEMMA 3.4 ([MAN 09a]).– *Let $\mathcal{H}$ be a hypothesis class, and let $\ell : Y \times Y \to \mathbb{R}_+$ define a loss function that is bounded, $\forall(y, y') \in Y^2$, $\ell(y, y') \leq M$ for some $M > 0$ and $L_{\mathcal{H}} = \{\mathbf{x} \to \ell(h'(\mathbf{x}), h(\mathbf{x})) : h, h' \in \mathcal{H}\}$. Let $\mathcal{D}_{\mathbf{X}}$ be a distribution over $\mathbf{X}$ and $\hat{\mathcal{D}}_{\mathbf{X}}$ denote the corresponding empirical distribution for a sample $S = (\mathbf{x}_1, \ldots, \mathbf{x}_m)$. Then, for any $\delta \in (0, 1)$, with probability at least $1 - \delta$ over the choice of sample S, we have*

$$disc_{\ell}(\mathcal{D}_{\mathbf{X}}, \hat{\mathcal{D}}_{\mathbf{X}}) \leq \mathcal{R}_S(L_{\mathcal{H}}) + 3M\sqrt{\frac{\log \frac{2}{\delta}}{2m}},$$

where $\mathcal{R}_S(L_{\mathcal{H}})$ is the empirical Rademacher complexity of $L_{\mathcal{H}}$ based on the observations from S.

The proof of this lemma can be found in Appendix 1. One may note that this lemma looks very much like the usual generalization inequalities obtained using the Rademacher complexities presented in section 1.2.2. Using this result, we can further prove the following corollary for the case of more general loss functions defined as $\forall(y, y') \in Y^2$, $\ell_q(y, y') = |y - y'|^q$ for some q. This parametric family of functions is a common choice of loss function for a regression task.

COROLLARY 3.1 ([MAN 09a]).– *Let $\mathcal{S}$ and $\mathcal{T}$ be the source and the target domain over $\mathbf{X} \times Y$, respectively. Let $\mathcal{H}$ be a hypothesis class and $\ell_q : Y \times Y \to \mathbb{R}_+$ be a loss function that is bounded, $\forall(y, y') \in Y^2$, $\ell_q(y, y') \leq M$ for some $M > 0$, and defined as $\forall(y, y') \in Y^2$, $\ell_q(y, y') = |y - y'|^q$ for some q. Let S_u and T_u be samples of size m_s and m_t drawn independently from $\mathcal{S}_{\mathbf{X}}$ and $\mathcal{T}_{\mathbf{X}}$. Denote by $\hat{\mathcal{S}}_{\mathbf{X}}, \hat{\mathcal{T}}_{\mathbf{X}}$ the empirical distributions corresponding to $\mathcal{S}_{\mathbf{X}}$ and $\mathcal{T}_{\mathbf{X}}$. Then, for any $\delta \in (0, 1)$, with probability at least $1 - \delta$ over the random choice of samples, we have*

$$\begin{aligned} disc_{\ell_q}(\mathcal{S}_{\mathbf{X}}, \mathcal{T}_{\mathbf{X}}) \leq\ & disc_{\ell_q}(\hat{\mathcal{S}}_{\mathbf{X}}, \hat{\mathcal{T}}_{\mathbf{X}}) + 4q\left(\mathcal{R}_{S_u}(\mathcal{H}) + \mathcal{R}_{T_u}(\mathcal{H})\right) \\ & +3M\left(\sqrt{\frac{\log(\frac{4}{\delta})}{2m_s}} + \sqrt{\frac{\log(\frac{4}{\delta})}{2m_t}}\right). \end{aligned}$$

The proof of this corollary can be found in Appendix 1. This result highlights one of the major differences between the approach of section 3.1 [BEN 10a] and that of Mansour *et al.* [MAN 09a] that lies in the way they estimate the introduced distance. While theorem 3.2 relies on VC dimension to bound the true $\mathcal{H}\Delta\mathcal{H}$-divergence by its empirical counterpart, $disc_\ell$ is estimated using the quantities based on the Rademacher complexity. In order to illustrate what it implies for the generalization guarantees, we now present the analog of theorem 3.2 that relates the source and target error function using the discrepancy distance and compare it to the original result.

THEOREM 3.4 ([MAN 09a]).– *Let $\mathcal{S}$ and $\mathcal{T}$ be the source and the target domain over $\mathbf{X} \times Y$, respectively. Let $\mathcal{H}$ be a hypothesis class; $\ell : Y \times Y \to \mathbb{R}_+$ be a loss function that is symmetric, obeys the triangle inequality and is bounded; $\forall (y, y') \in Y^2$; and $\ell(y, y') \leq M$ for some $M > 0$. Then, for $h^*_\mathcal{S} = \underset{h \in \mathcal{H}}{\operatorname{argmin}} \ \mathrm{R}^\ell_\mathcal{S}(h)$ and $h^*_\mathcal{T} = \underset{h \in \mathcal{H}}{\operatorname{argmin}} \ \mathrm{R}^\ell_\mathcal{T}(h)$ denoting the ideal hypotheses for the source and target domains, we have*

$$\forall h \in \mathcal{H}, \ \mathrm{R}^\ell_\mathcal{T}(h) \leq \mathrm{R}^\ell_\mathcal{S}(h, h^*_\mathcal{S}) + disc_\ell(\mathcal{S}_\mathbf{X}, \mathcal{T}_\mathbf{X}) + \epsilon,$$

$$\textit{where } \mathrm{R}^\ell_\mathcal{S}(h, h^*_\mathcal{S}) = \underset{\mathbf{x} \sim \mathcal{S}_\mathbf{X}}{\mathbf{E}} \ \ell\left(h(\mathbf{x}), h^*_\mathcal{S}(\mathbf{x})\right) \textit{ and } \epsilon = \mathrm{R}^\ell_\mathcal{T}(h^*_\mathcal{T}) + \mathrm{R}^\ell_\mathcal{S}(h^*_\mathcal{T}, h^*_\mathcal{S}).$$

The proof of this theorem can be found in Appendix 1. As pointed out by the authors, this bound is not directly comparable to theorem 3.2, but involves similar terms and reflects a very common trade-off between them. Indeed, the first term of this bound stands for the same source risk function as the one found in [BEN 10a]. The second term here captures the deviation between the two domains through the discrepancy distance similar to the $\mathcal{H}\Delta\mathcal{H}$-divergence used before. Finally, the last term ϵ can be interpreted as the capacity to adapt and is very close in spirit to the λ term seen previously. Consequently, this result suggests the same algorithmic implications as that spurred by theorem 3.2: assuming that ϵ is low, an efficient adaptation algorithm has to find a good representation space where the two domains are close, while keeping good performance in the source domain.

Despite these similarities, a closer comparison made in [MAN 09a] revealed that the bound based on the discrepancy distance can be more tight in some plausible scenarios. For instance, in a degenerate case where there is only one hypothesis $h \in \mathcal{H}$ and a single target function $f_\mathcal{T}$, the bounds of theorem 3.4 and of theorem 3.2 with true distributions give $\mathrm{R}^\ell_\mathcal{T}(h, f) + disc_\ell(\mathcal{S}_\mathbf{X}, \mathcal{T}_\mathbf{X})$ and $\mathrm{R}^\ell_\mathcal{T}(h, f) + 2\mathrm{R}^\ell_\mathcal{S}(h, f) + disc_\ell(\mathcal{S}_\mathbf{X}, \mathcal{T}_\mathbf{X})$, respectively. In this case, the latter expression is obviously larger when $\mathrm{R}^\ell_\mathcal{S}(h, f) \leq \mathrm{R}^\ell_\mathcal{T}(h, f)$. The same kind of result can be also shown to hold under the following plausible assumptions:

1) when $h^* = h^*_\mathcal{S} = h^*_\mathcal{T}$, the bounds of theorems 3.4 and 3.2, respectively, boil down to

$$\mathrm{R}^\ell_\mathcal{T}(h) \leq \mathrm{R}^\ell_\mathcal{T}(h^*) + \mathrm{R}^\ell_\mathcal{S}(h, h^*) + disc_\ell(\mathcal{S}_\mathbf{X}, \mathcal{T}_\mathbf{X}), \qquad [3.3]$$

and

$$\mathrm{R}^\ell_\mathcal{T}(h) \leq \mathrm{R}^\ell_\mathcal{T}(h^*) + \mathrm{R}^\ell_\mathcal{S}(h^*) + \mathrm{R}^\ell_\mathcal{S}(h) + disc_\ell(\mathcal{S}_\mathbf{X}, \mathcal{T}_\mathbf{X}), \qquad [3.4]$$

where the right-hand side of equation [3.4] essentially includes the sum of three errors and is always larger than the right-hand side of equation [3.3] due to the triangle inequality.

2) when $h^* = h^*_{\mathcal{S}} = h^*_{\mathcal{T}}$ and $disc_\ell(\mathcal{S}_\mathbf{X}, \mathcal{T}_\mathbf{X}) = 0$, theorems 3.4 and 3.2 give

$$\mathrm{R}^\ell_{\mathcal{T}}(h) \leq \mathrm{R}^\ell_{\mathcal{T}}(h^*) + \mathrm{R}^\ell_{\mathcal{S}}(h, h^*)$$

and

$$\mathrm{R}^\ell_{\mathcal{T}}(h) \leq \mathrm{R}^\ell_{\mathcal{T}}(h^*) + \mathrm{R}^\ell_{\mathcal{S}}(h^*) + \mathrm{R}^\ell_{\mathcal{S}}(h),$$

where the former coincides with the standard generalization bound while the latter does not.

3) finally, when $f_{\mathcal{T}} \in \mathcal{H}$, theorem 3.2 simplifies to

$$|\mathrm{R}^\ell_{\mathcal{T}}(h) - \mathrm{R}^\ell_{\mathcal{S}}(h)| \leq disc_{\ell_{01}}(\mathcal{S}_\mathbf{X}, \mathcal{T}_\mathbf{X}),$$

which can be straightforwardly obtained from theorem 3.4.

All these results show a tight link that can be observed in different contributions of the domain adaptation theory. This relation essentially illustrates that the results of [MAN 09a] strengthen the previous results on the subject but keep a tight connection to them.

3.5.3. *Domain adaptation in regression*

As mentioned in the beginning of this section, the discrepancy distance not only extends the first theoretical results obtained for domain adaptation, but also allows us to derive new point-wise guarantees for other learning scenarios such as the regression task where, contrary to classification, the output variable Y is continuous. The domain adaptation problem for the regression task can be illustrated as in Figure 3.3.

To this end, another type of theoretical result based on the discrepancy distance was proposed by Cortes and Mohri [COR 11] for the regression problem. The authors consider the case where the hypothesis set $\mathcal{H}$ is a subset of the reproducing kernel Hilbert space (RKHS) $\mathbb{H}$ associated with a positive definite symmetric (PDS) kernel $K : \mathcal{H} = \{h \in \mathbb{H} : \|h\|_K \leq \Lambda\}$, where $\|\cdot\|_K$ denotes the norm defined by the inner product on $\mathbb{H}$ and $\Lambda \geq 0$. We shall assume that there exists $R > 0$ such that $K(\mathbf{x}, \mathbf{x}) \leq R^2$ for all $\mathbf{x} \in \mathbf{X}$. By the reproducing property, for any $h \in \mathcal{H}$ and $\mathbf{x} \in \mathbf{X}$, $h(\mathbf{x}) = \langle h, K(\mathbf{x}, \cdot)\rangle_K$, and thus this implies that $|h(\mathbf{x})| \leq \|h\|_K \sqrt{K(\mathbf{x}, \mathbf{x})} \leq \Lambda R$.

In this setting, the authors further present point-wise loss guarantees in domain adaptation for a broad class of kernel-based regularization algorithms. Given a learning sample S, where $\forall(\mathbf{x}, y) \in S, \mathbf{x} \sim \mathcal{D}_{\mathbf{X}},\ y = f_{\mathcal{D}}(\mathbf{x})$, these algorithms are defined by the minimization of the following objective function:

$$F_{\hat{\mathcal{D}}_{\mathbf{X}}}(h) = \mathrm{R}^{\ell}_{\hat{\mathcal{D}}_{\mathbf{X}}}(h, f_{\mathcal{D}}) + \beta \|h\|^2_K,$$

where $\beta > 0$ is a trade-off parameter. This family of algorithms includes SVM, support vector regression (SVR) [VAP 95], kernel ridge regression (KRR) [SAU 98] and many other methods. Finally, the loss function ℓ is also assumed to be μ-admissible following the definition given below.

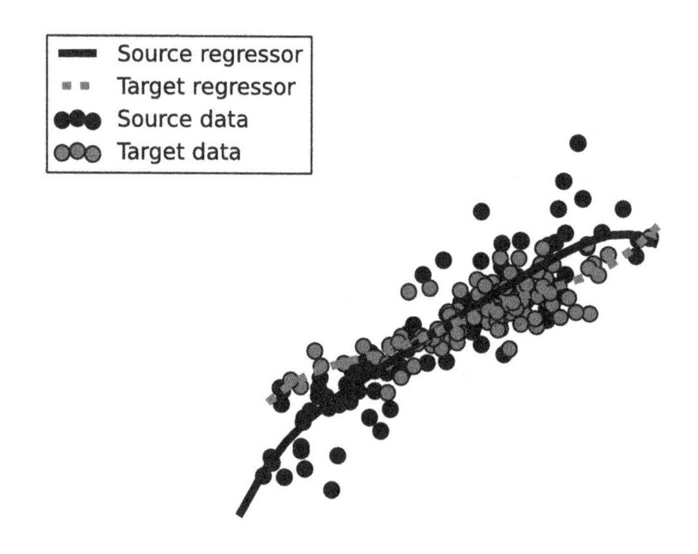

Figure 3.3. *Regression problem in domain adaptation context. For a color version of this figure, see www.iste.co.uk/redko/domain.zip*

DEFINITION 3.4 (μ-admissible loss).– *A loss function $\ell : Y \times Y \to \mathbb{R}$ is μ-admissible if it is symmetric and convex with respect to both of its arguments, and for all $\mathbf{x} \in \mathbf{X}$, $y \in Y$ and $(h, h') \in \mathcal{H}^2$, it verifies the following Lipschitz condition for some $\mu > 0$:*

$$|\ell(h'(\mathbf{x}), y) - \ell(h(\mathbf{x}), y)| \leq \mu |h'(\mathbf{x}) - h(\mathbf{x})|.$$

The family of μ-admissible losses includes the hinge loss and all $\ell_q(y, y') = |y - y'|^q$ with $q \geq 1$, in particular the squared loss, when the hypothesis set and the set of output labels are bounded.

With the assumptions made previously, the following results can be proved.

THEOREM 3.5 ([COR 11, COR 14]).– *Let $\mathcal{S}$ and $\mathcal{T}$ be the source and the target domain on $\mathbf{X} \times Y$, $\mathcal{H}$ be a hypothesis class and ℓ be a μ-admissible loss. We assume that the target labeling function $f_{\mathcal{T}}$ belongs to $\mathcal{H}$, and let η denote $\max\{\ell(f_{\mathcal{S}}(\mathbf{x}), f_{\mathcal{T}}(\mathbf{x})) : \mathbf{x} \in \text{SUPP}(\hat{\mathcal{S}}_{\mathbf{X}})\}$. Let h' be the hypothesis minimizing $F_{\hat{\mathcal{T}}_{\mathbf{X}}}$ and h the one returned when minimizing $F_{\hat{\mathcal{S}}_{\mathbf{X}}}$. Then, for all $(\mathbf{x}, y) \in \mathbf{X} \times Y$, we have*

$$|\ell(h'(\mathbf{x}), y) - \ell(h(\mathbf{x}), y)| \leq \mu R \sqrt{\frac{disc_\ell(\hat{\mathcal{S}}_{\mathbf{X}}, \hat{\mathcal{T}}_{\mathbf{X}}) + \mu\eta}{\beta}}.$$

The proof of this theorem can be found in Appendix 1. This theorem essentially shows that the difference between the errors achieved by optimal hypotheses learned on source and target samples is proportional to the distance between samples plus a term reflecting the worst value that a loss function can achieve for some instance belonging to the support of $\hat{\mathcal{S}}_{\mathbf{X}}$.

A similar theorem can be proven when not $f_{\mathcal{T}}$ but $f_{\mathcal{S}} \in \mathcal{H}$ is assumed. Furthermore, the authors mention that these theorems can be extended to the case where neither the target function $f_{\mathcal{T}}$ nor $f_{\mathcal{S}}$ belong to $\mathcal{H}$ by replacing η in the statement of the theorem with

$$\eta' = \max_{\mathbf{x} \in \text{SUPP}(\hat{\mathcal{S}}_{\mathbf{X}})} \{\ell(h^*_{\mathcal{T}}(\mathbf{x}), f_{\mathcal{S}}(\mathbf{x}))\} + \max_{\mathbf{x} \in \text{SUPP}(\hat{\mathcal{T}}_{\mathbf{X}})} \{\ell(h^*_{\mathcal{T}}(\mathbf{x}), f_{\mathcal{T}}(\mathbf{x}))\},$$

where $h^*_{\mathcal{T}} \in \underset{h \in \mathcal{H}}{\text{argmin}}\ \ell(h(\mathbf{x}), f_{\mathcal{T}})$. In both cases, when η is assumed to be small, i.e. $\eta \ll 1$, the key term of the obtained bound is the empirical discrepancy distance $disc_\ell(\hat{\mathcal{S}}_{\mathbf{X}}, \hat{\mathcal{T}}_{\mathbf{X}})$. In the extreme case where $f_{\mathcal{T}} = f_{\mathcal{S}} = f$, we obtain $\eta = 0$ and the problem reduces to the covariate shift adaptation scenario described in section 2.2.1. In general, one can draw a parallel between the η term that appears in this bound and the other so-called adaptation capacity terms as the λ term in the bound of Ben-David *et al.* from theorem 3.2. Once again, it shows that efficient adaptation cannot entirely depend on the distance between the unlabeled samples of source and target domains.

From a practical point of view, under the assumptions $\eta \ll 1$ or $\eta' \ll 1$, these theorems suggest seeking an empirical distribution q^*, among the family $\mathcal{Q}$ of all distributions with a support included in that of $\hat{\mathcal{S}}_{\mathbf{X}}$, that minimizes that discrepancy:

$$q^* = \underset{q \in \mathcal{Q}}{\text{argmin}}\ disc_\ell(\hat{\mathcal{T}}_{\mathbf{X}}, q).$$

As mentioned in the original paper [COR 11], using q^* instead of $\hat{\mathcal{S}}_{\mathbf{X}}$ amounts to re-weighting the loss on each training point. This allows us to derive the basis of an adaptation algorithm which consists of:

1) first computing q^*;

2) then modifying the optimized criterion using q^*:

$$\min_{h\in\mathcal{H}} \frac{1}{m}\sum_{i=1}^{m} q^*(\mathbf{x}_i)\ell(h(\mathbf{x}_i), y_i) + \beta\|h\|_K^2,$$

and finding a minimizing hypothesis h.

Solving this problem presents no particular challenge compared to the classical minimization of the empirical risk over a given sample.

The result given by theorem 3.5 can be further strengthened when the considered loss function is assumed to be squared loss $\ell_2 := (y - y')^2$ for some $(y, y') \in Y^2$ and when the kernel-based regularization algorithm described above coincides with the KRR. In what follows, the term η will be replaced by a finer quantity defined as

$$\delta_{\mathcal{H}}(f_S, f_T) = \inf_{h\in\mathcal{H}} \| \mathop{\mathbf{E}}_{\mathbf{x}\sim\hat{\mathcal{S}}_{\mathbf{X}}} [\Delta(h, f_S)] - \mathop{\mathbf{E}}_{\mathbf{x}\sim\hat{\mathcal{T}}_{\mathbf{X}}} [\Delta(h, f_T)] \|,$$

where $\Delta(h, f) = (f(\mathbf{x}) - h(\mathbf{x}))\Phi(\mathbf{x})$ with $\Phi(\mathbf{x})$ being associated with kernel K feature vector so that $K(\mathbf{x}, \mathbf{x}') = \langle\Phi(\mathbf{x}), \Phi(\mathbf{x}')\rangle$.

Using this quantity, the following guarantee holds.

THEOREM 3.6 ([COR 14]).– *Let ℓ be a squared loss bounded by some $M > 0$ and let h' be the hypothesis minimizing $F_{\hat{\mathcal{T}}_{\mathbf{X}}}$ and h the one returned when minimizing $F_{\hat{\mathcal{S}}_{\mathbf{X}}}$. Then, for all $(\mathbf{x}, y) \in \mathbf{X} \times Y$, we have:*

$$|\ell(h(\mathbf{x}), y) - \ell(h'(\mathbf{x}), y)|$$

$$\leq \frac{R\sqrt{M}}{\beta}\left(\delta_{\mathcal{H}}(f_S, f_T) + \sqrt{\delta_{\mathcal{H}}^2(f_S, f_T) + 4\,\beta\, disc_\ell(\hat{\mathcal{S}}_{\mathbf{X}}, \hat{\mathcal{T}}_{\mathbf{X}})}\right).$$

The proof of this theorem can be found in Appendix 1. As pointed out by the authors, the main advantage of this result is its expression in terms of $\delta_{\mathcal{H}}(f_S, f_T)$ instead of $\eta_{\mathcal{H}}(f_S, f_T)$. One may note that $\delta_{\mathcal{H}}(f_S, f_T)$ is defined as a difference and thus it becomes zero for $\mathcal{S}_{\mathbf{X}} = \mathcal{T}_{\mathbf{X}}$, which does not hold for $\eta_{\mathcal{H}}(f_S, f_T)$. Furthermore, when the covariate-shift assumption holds for some shared labeling function f such that $f_S = f_T = f$, $\delta_{\mathcal{H}}(f, f)$ can be upper bounded using the following result.

THEOREM 3.7 ([COR 14]).– *Assume that for all $\mathbf{x} \in \mathbf{X}$, $K(\mathbf{x}, \mathbf{x}) \leq R^2$ for some $R > 0$. Let $\mathcal{A}$ denote the union of the supports of $\hat{\mathcal{S}}_{\mathbf{X}}$ and $\hat{\mathcal{T}}_{\mathbf{X}}$. Then, for any $p > 1$ and $q > 1$, with $1/p + 1/q = 1$,*

$$\delta_{\mathcal{H}}(f, f) \leq d_p(f_{|\mathcal{A}}, \mathcal{H}_{|\mathcal{A}})\ell_q(\hat{\mathcal{S}}_{\mathbf{X}}, \hat{\mathcal{T}}_{\mathbf{X}}),$$

where for any set $\mathcal{A} \subseteq \mathbf{X}$, $f_{|\mathcal{A}}$ (respectively, $\mathcal{H}_{|\mathcal{A}}$) denote the restriction of f (respectively, h) to $\mathcal{A}$ and $d_p(f_{|\mathcal{A}}, \mathcal{H}_{|\mathcal{A}}) = \inf_{h\in\mathcal{H}} \|f - h\|_p$.

The proof of this theorem can be found in Appendix 1. In particular, the authors show that for a labeling function f that belongs to the closure of $\mathcal{H}_{|\mathcal{A}}$, $\delta_{\mathcal{H}}(f) = 0$ when the KRR algorithm is used with normalized Gaussian kernels. For this specific algorithm, which often deployed in practice, the bound of the theorem then reduces to the simpler expression:

$$|\ell(h(\mathbf{x}), y) - \ell(h'(\mathbf{x}), y)| \leq 2R\sqrt{\frac{M disc_\ell(\hat{\mathcal{S}}_\mathbf{X}, \hat{\mathcal{T}}_\mathbf{X})}{\beta}}.$$

The results mentioned above can be further strengthened using a recently proposed notation of the generalized discrepancy introduced by Cortes *et al.* [COR 15]. In order to introduce this distance, we may first note that a regression task in the domain adaptation context can be seen as an optimal approximation of an ideal hypothesis $h^*_\mathcal{T} = \underset{h \in \mathcal{H}}{\text{argmin}}\ \mathrm{R}^\ell_\mathcal{T}(h, f_\mathcal{T})$ by another hypothesis h that ensures the closeness of the losses $\mathrm{R}^\ell_\mathcal{T}(h^*, f_\mathcal{T})$ and $\mathrm{R}^\ell_\mathcal{T}(h, f_\mathcal{T})$. As we do not have access to $f_\mathcal{T}$ but only to the labels of the source sample S, the main idea is to define, for any $h \in \mathcal{H}$, a reweighting function $Q_h : S \to \mathbb{R}$ such that the objective function G defined for all $h \in \mathcal{H}$ by

$$G(h) = \mathrm{R}^\ell_{Q_h}(h) + \beta\|h\|^2_K,$$

remains uniformly close to $F_{\hat{\mathcal{T}}_\mathbf{X}}(h)$ defined over the target sample T_u. As pointed out by the authors, this idea introduces a different learning concept as instead of reweighting the training sample with some fixed set of weights, the weights are allowed to vary as a function of the hypothesis h and are not assumed to sum to one or to be non-negative. Based on this construction, the optimal reweighting can be obtained by solving:

$$Q_h = \underset{q \in \mathcal{F}(\mathcal{S}_\mathbf{X}, \mathbb{R})}{\text{argmin}} |\mathrm{R}^\ell_{\hat{\mathcal{T}}_\mathbf{X}}(h, f_\mathcal{T}) - \mathrm{R}^\ell_q(h, f_S)|,$$

where $\mathcal{F}(\mathcal{S}_\mathbf{X}, \mathbb{R})$ is the set of real-valued functions defined over $\textsc{supp}(\mathcal{S}_\mathbf{X})$.

We can note that, in practice, we may not have access to target labeled samples, which implies that we cannot estimate $f_\mathcal{T}$. To solve this problem, the authors propose to consider a non-empty convex set of candidate hypotheses $\mathcal{H}'' \subseteq \mathcal{H}$ that could contain a good approximation of $f_\mathcal{T}$. Using $\mathcal{H}''$ as a set of surrogate labeling functions, the previous optimization problem becomes:

$$Q_h = \underset{q \in \mathcal{F}(\mathcal{S}_\mathbf{X}, \mathbb{R})}{\text{argmin}} \max_{h'' \in \mathcal{H}} |\mathrm{R}^\ell_{\hat{\mathcal{T}}_\mathbf{X}}(h, h'') - \mathrm{R}^\ell_q(h, f_S)|.$$

The risk obtained using the solution of this optimization problem given by Q_h can be equivalently expressed as follows:

$$\mathrm{R}^\ell_{Q_h}(h, f_S) = \frac{1}{2}\left(\max_{h'' \in \mathcal{H}} \mathrm{R}^\ell_{\hat{\mathcal{T}}_\mathbf{X}}(h, h'') + \min_{h'' \in \mathcal{H}} \mathrm{R}^\ell_{\hat{\mathcal{T}}_\mathbf{X}}(h, h'')\right).$$

This last expression was obtained by the authors based on the following reasoning. For any $h \in \mathcal{H}$, the equation $\mathrm{R}^{\ell}_{Q_h}(h, f_S) = l$ with $l \in \mathbb{R}$ admits a solution $q \in \mathcal{F}(\mathcal{S}_{\mathbf{X}}, \mathbb{R})$. Consequently, we can write $\forall h \in \mathcal{H}$:

$$\begin{aligned}
\mathrm{R}^{\ell}_{Q_h}(h, f_S) &= \underset{l \in \ell_q(h, f_S): q \in \mathcal{F}(\mathcal{S}_{\mathbf{X}}, \mathbb{R})}{\mathrm{argmin}} \max_{h, h'' \in \mathcal{H}} \left| l - \mathrm{R}^{\ell}_{\hat{\mathcal{T}}_{\mathbf{X}}}(h, h'') \right| \\
&= \underset{l \in \mathbb{R}}{\mathrm{argmin}} \max_{h, h'' \in \mathcal{H}} \left| l - \mathrm{R}^{\ell}_{\hat{\mathcal{T}}_{\mathbf{X}}}(h, h'') \right| \\
&= \underset{l \in \mathbb{R}}{\mathrm{argmin}} \max_{h, h'' \in \mathcal{H}} \max\{\mathrm{R}^{\ell}_{\hat{\mathcal{T}}_{\mathbf{X}}}(h, h'') - l, l - \mathrm{R}^{\ell}_{\hat{\mathcal{T}}_{\mathbf{X}}}(h, h'')\} \\
&= \underset{l \in \mathbb{R}}{\mathrm{argmin}} \max\{\max_{h, h'' \in \mathcal{H}} \mathrm{R}^{\ell}_{\hat{\mathcal{T}}_{\mathbf{X}}}(h, h'') - l, l - \min_{h, h'' \in \mathcal{H}} \mathrm{R}^{\ell}_{\hat{\mathcal{T}}_{\mathbf{X}}}(h, h'')\} \\
&= \frac{1}{2} \left(\max_{h'' \in \mathcal{H}} \mathrm{R}^{\ell}_{\hat{\mathcal{T}}_{\mathbf{X}}}(h, h'') + \min_{h'' \in \mathcal{H}} \mathrm{R}^{\ell}_{\hat{\mathcal{T}}_{\mathbf{X}}}(h, h'') \right).
\end{aligned}$$

This, in its turn, allows us to reformulate $G(h)$, which now becomes:

$$G(h) = \frac{1}{2} \left(\max_{h'' \in \mathcal{H}} \mathrm{R}^{\ell}_{\hat{\mathcal{T}}_{\mathbf{X}}}(h, h'') + \min_{h'' \in \mathcal{H}} \mathrm{R}^{\ell}_{\hat{\mathcal{T}}_{\mathbf{X}}}(h, h'') \right) + \beta \|h\|^2_K.$$

The proposed optimization problem should have the same point-wise guarantees as those established in theorem 3.5.3. Contrary to the latter results, the learning bounds for this approach rely on a new notation of the distance between the probability distributions that can be seen as a generalization of the discrepancy distance used before. In order to introduce it, we now define $A(\mathcal{H})$ as a set of functions $U : h \to U_h$ that map $\mathcal{H}$ to $\mathcal{F}(\mathcal{S}_{\mathbf{X}}, \mathbb{R})$ such that for all $h \in \mathcal{H}$, $h \to \ell_{U_h}(h, f_S)$ is a convex function. The set $A(\mathcal{H})$ contains all constant functions U such that $U_h = q$ for all $h \in \mathcal{H}$, where q is a distribution over $\mathcal{S}_{\mathbf{X}}$. The definition of the generalized discrepancy can thus be given as follows.

DEFINITION 3.5.– *For any $U \in A(\mathcal{H})$, the generalized discrepancy between U and $\hat{\mathcal{T}}_{\mathbf{X}}$ is defined as*

$$DISC(\hat{\mathcal{T}}_{\mathbf{X}}, U) = \max_{h \in \mathcal{H}, h'' \in \mathcal{H}''} \left| \mathrm{R}^{\ell}_{\hat{\mathcal{T}}_{\mathbf{X}}}(h, h'') - \mathrm{R}^{\ell}_{U_h}(h, f_S) \right|.$$

In addition, the authors also defined the following distance of f to $\mathcal{H}''$ over the support of $\hat{\mathcal{T}}_{\mathbf{X}}$:

$$d^{\hat{\mathcal{T}}_{\mathbf{X}}}_{\infty}(f_{\mathcal{T}}, \mathcal{H}'') = \min_{h_0 \in \mathcal{H}''} \max_{\mathbf{x} \in \mathrm{SUPP}(\hat{\mathcal{T}}_{\mathbf{X}})} |h_0(\mathbf{x}) - f_{\mathcal{T}}(\mathbf{x})|.$$

Using the above-defined quantities, the following point-wise guarantees can be given.

THEOREM 3.8 ([COR 15]).– *Let* h^* *be a minimizer of* $\mathrm{R}^{\ell}_{\hat{\mathcal{T}}_{\mathbf{X}}}(h, f_{\mathcal{T}}) + \beta\|h\|^2_K$ *and* h_Q *be a minimizer of* $\mathrm{R}^{\ell}_{Q_h}(h, f_S) + \beta\|h\|^2_K$*. Then, for* $Q : h \to Q_h$ *and* $\forall \mathbf{x} \in \mathbf{X},\ y \in Y$*, the following holds:*

$$|\ell(h_Q(\mathbf{x}), y) - \ell(h^*(\mathbf{x}), y)| \leq \mu R \sqrt{\frac{\mu d_{\infty}^{\hat{\mathcal{T}}_{\mathbf{X}}}(f_{\mathcal{T}}, \mathcal{H}'') + DISC(Q, \hat{\mathcal{T}}_{\mathbf{X}})}{\beta}}.$$

Furthermore, this inequality can equivalently be written in terms of the risk functions as

$$\mathrm{R}^{\ell}_{\mathcal{T}}(h_{\mathrm{Q}}, f_{\mathcal{T}}) \leq \mathrm{R}^{\ell}_{\mathcal{T}}(h^*, f_{\mathcal{T}}) + \mu R \sqrt{\frac{\mu d_{\infty}^{\hat{\mathcal{T}}_{\mathbf{X}}}(f_{\mathcal{T}}, \mathcal{H}'') + \mathrm{DISC}(\mathrm{Q}, \hat{\mathcal{T}}_{\mathbf{X}})}{\beta}}.$$

The proof of this theorem can be found in Appendix 1. The result of theorem 3.8 suggests selecting $\mathcal{H}''$ to minimize the right-hand side of the last inequality. In particular, the authors provide further evidence that if the space over which $\mathcal{H}''$ is searched is the family of all balls centered in f_S defined in terms of l_{q^*}, i.e. $\mathcal{H}'' = \{h'' \in \mathcal{H} | l_q(h'', f_Q) \leq r\}$ for some distribution q over the space of the reweighted source samples, then the proposed algorithm based on the generalized discrepancy gives provably better results compared to the original one. Furthermore, in a semi-supervised scenario where some labeled data in the target domain are available, we can actually use part of it to find an appropriate value of r. In order to prove this latter statement, let us consider a situation where in addition to the unlabeled target sample T_u and the labeled source sample S the learner also receives a small labeled sample T. Then, we consider the following set $S' = S \cup T$ and an empirical distribution $\hat{\mathcal{S}}'_{\mathbf{X}}$ over it.

Let us denote by q'^* the distribution minimizing the discrepancy between $\hat{\mathcal{S}}'_{\mathbf{X}}$ and $\hat{\mathcal{T}}_{\mathbf{X}}$. Now since the support $\textsc{supp}(\hat{\mathcal{S}}_{\mathbf{X}})$ is in that of $\textsc{supp}(\hat{\mathcal{S}}'_{\mathbf{X}})$, the following inequality can be obtained:

$$\begin{aligned} disc_{\ell}(\hat{\mathcal{T}}_{\mathbf{X}}, q'^*) &= \min_{\mathrm{SUPP}(q) \subseteq \mathrm{SUPP}(\hat{\mathcal{S}}'_{\mathbf{X}})} disc_{\ell}(\hat{\mathcal{T}}_{\mathbf{X}}, q) \\ &\leq \min_{\mathrm{SUPP}(q) \subseteq \mathrm{SUPP}(\hat{\mathcal{S}}_{\mathbf{X}})} disc_{\ell}(\hat{\mathcal{T}}_{\mathbf{X}}, q) = disc_{\ell}(\hat{\mathcal{T}}_{\mathbf{X}}, q^*). \end{aligned}$$

Consequently, in view of theorem 3.8, for an appropriate choice of $\mathcal{H}''$, the learning guarantee for adaptation algorithms based on the generalized discrepancy is more favorable when using some labeled data from the target domain. Thus, using the limited number of labeled points from the target distribution can improve the performance of their proposed algorithm.

3.6. Summary

This chapter presents several cornerstone results of the domain adaptation theory, including those proposed by Ben-David *et al.* based on the $\mathcal{H}\Delta\mathcal{H}$-divergence and a variety of results based on the discrepancy distance proposed by Mansour *et al.* and Cortes *et al.* for the tasks of classification and regression. As one may note, the general ideas used to prove generalization bounds for domain adaptation are based on:

1) the definition of a relation between the source and the target domain through a divergence allowing us to upper bound the target risk by the source risk;

2) the theoretical results presented in Chapter 1 and thus inherit all their well-known properties.

Unsurprisingly, this trend is usually maintained regardless the considered domain adaptation scenario and analyzed learning algorithms. The overall form of the presented generalization bound on the error of a hypothesis calculated with respect to the target distribution inevitably appears to contain the following important terms:

1) The source error of the hypothesis measured with respect to some loss function.

2) The divergence term between marginal distributions of source and target domains. In the case of Ben-David *et al.*, this term is explicitly linked to the hypothesis space inducing a complexity term related to its Vapnik–Chervonenkis dimension; in the case of Mansour *et al.* and Cortes *et al.*, the divergence term depends on the hypothesis space but the complexity term is data dependent and is linked to the Rademacher complexity of the hypothesis space.

3) The non-estimable term that reflects the *a priori* hardness of the domain adaptation problem. This latter usually requires at least some target labeled data in order to be quantified.

The first term from this list shows us that in the case where two domains are almost indistinguishable, the performance of a given hypothesis across them will remain largely similar. When this is not the case, the divergence between the source and target domain marginal distributions starts to play a crucial role in assessing the proximity of the two domains. For both presented results, the actual value of this divergence can be consistently calculated using the available finite (unlabeled) samples, thus providing us with a first estimate of the potential success of adaptation. Finally, the last term tells us that even bringing the divergence between the marginal distributions to zero across two domains may not suffice for efficient adaptation. Indeed, this result is rather intuitive as we can find a mapping that projects source and target data to the same space (or aligns one of them with another), but that mapping needs to take into account the unknown labeling of the source and target points in order to ensure efficient classification afterwards. The last point can be summarized by the following statement made by Ben-David [BEN 10a]:

> "When the combined error of the ideal joint hypothesis is large, then there is no classifier that performs well on both the source and target domains, so we cannot hope to find a good target hypothesis by training only on the source domain".

This statement brings us to another important question regarding the conditions that one needs to verify in order to make sure that the adaptation can be successful. This question spurs a cascade of other relevant inquiries, such as, what is the actual size of the source and target unlabeled samples needed for adaptation to be efficient? Are target labeled data needed for efficient adaptation and if yes can we prove formally that it leads to better results? And finally, what are the pitfalls of domain adaptation when even strong prior knowledge regarding the adaptation problem does not guarantee it to have a solution? All these question are answered by the so-called "impossibility theorems" that we present in the following chapter.

4

Impossibility Theorems for Domain Adaptation

In many scientific areas, we often seek to obtain general conclusions about the behavior of an object of interest and different phenomena related to its behavior. The conclusions that we are usually aiming for mainly concern the conditions under which these phenomena occur as well as assumptions that prevent them from happening. In mathematics, for instance, a statement can be proved or rejected: while first result is obviously what one is looking for, the second is also extremely valuable for understanding the studied properties of an object. A very direct analogy to this can also be drawn regarding the main subject of this book. In the previous chapters, we have already reviewed the problem of domain adaptation from the theoretical point of view in a variety of different paradigms. This analysis provided us with strong evidence about the necessary conditions that lead to efficient adaptation. They include (i) proper discrepancy minimization between the source and the target domains and (ii) the existence of a good hypothesis that leads to a low error for two domains. These conditions are very important because they give us a thorough theoretical understanding of what domain adaptation is and guide us in designing new effective adaptation algorithms.

Since the main purpose of this book is to present an in-depth theoretical study of domain adaptation, we cannot ignore the problem of understanding whether derived conditions are really necessary or whether they can be replaced or extended using a new set of assumptions. Thus, this chapter is devoted to a series of results that prove the so-called "impossibility theorems": statements showing that a certain set of assumptions does not ensure good adaptation. These theorems are very important as they provably show that in some cases one cannot hope to adapt well even with a prohibitively large amount of data from both domains.

4.1. Problem set-up

Before presenting the main theoretical contributions of this chapter, we first introduce the necessary preliminary definitions that formalize the concepts used in the impossibility theorems. These definitions are then followed by a set of assumptions that are commonly considered to have a direct influence on the potential success of domain adaptation.

4.1.1. *Definitions*

We have seen from previous chapters that adaptation efficiency is directly correlated with two main terms that inevitably appear in almost all analyses: one term shows the divergence between the domains while the other one stands for the existence and the error achieved by the best hypothesis across the source and target domains. Ben-David *et al.* [BEN 10b] propose to analyze the presence of these two terms in the bounds by answering the following questions:

– is the presence of these two terms inevitable in the domain adaptation bounds?

– is there a way to design a more intelligent domain adaptation algorithm that uses not only the labeled training sample but also the unlabeled sample of the target data distribution?

These two questions are very important as answering them can help us to obtain an exhaustive set of conditions that theoretically ensure efficient adaptation with respect to a given domain adaptation algorithm. Before proceeding to the presentation of the main results, the authors first define several quantities that are used later. The first one is the formalization of a unsupervised domain adaptation algorithm.

DEFINITION 4.1 (Domain adaptation learner).– *A domain adaptation learner is a function:*

$$\mathcal{A} : \bigcup_{m=1}^{\infty} \bigcup_{n=1}^{\infty} (\mathbf{X} \times \{0,1\})^m \times \mathbf{X}^n \to \{0,1\}^{\mathbf{X}}.$$

As mentioned before, the standard notation for the performance of the learner is given by the used error function. When the error is measured with respect to the best hypothesis in some hypothesis class $\mathcal{H}$, we use the notation $\mathrm{R}_{\mathcal{D}}(\mathcal{H}) = \inf_{h \in \mathcal{H}} \mathrm{R}_{\mathcal{D}}(h)$. Using this notation, the authors further define the learnability as follows.

DEFINITION 4.2 (Learnability).– *Let $\mathcal{S}$ and $\mathcal{T}$ be distributions over $\mathbf{X} \times \{0,1\}$, $\mathcal{H}$ a hypothesis class, $\mathcal{A}$ a domain adaptation learner, $\varepsilon > 0$, $\delta > 0$, and m, n positive integers. We say that $\mathcal{A}$ $(\varepsilon, \delta, m, n)$-learns $\mathcal{T}$ from $\mathcal{S}$ relative to $\mathcal{H}$ if, when given access to a labeled sample S of size m, generated i.i.d. by $\mathcal{S}$, and an unlabeled sample T_u*

of size n, generated i.i.d. by $\mathcal{T}_{\mathbf{X}}$, with probability at least $1 - \delta$ (over the choice of the samples S and T_u), the learned classifier does not exceed the P-error of the best classifier in $\mathcal{H}$ by more than ε, i.e.

$$\Pr_{\substack{S \sim (\mathcal{S})^m \\ T_u \sim (\mathcal{T}_{\mathbf{X}})^n}} \left[\mathrm{R}_{\mathcal{T}}(\mathcal{A}(S, T_u)) \leq \mathrm{R}_{\mathcal{T}}(\mathcal{H}) + \varepsilon \right] \geq 1 - \delta \, .$$

This definition gives us a criterion that we can use in order to judge if a particular algorithm has strong learning guarantees. It essentially means that for a good learnability, a given algorithm should efficiently minimize the trade-off between both ε and δ in the above definition.

4.1.2. *Common assumptions in domain adaptation*

We now present the most common assumptions that were considered in the literature as those that can ensure efficient adaptation. Most of them can be found in the previous chapters of this book, but for the sake of clarity we recall them all here.

Covariate shift. This assumption is among the most popular ones in domain adaptation and has been extensively studied in a series of theoretical works on the subject (see, for instance, [SUG 08] and the references therein). While in domain adaptation we generally assume $\mathcal{S} \neq \mathcal{T}$, this can be further understood as $\mathcal{S}_{\mathbf{X}}(\mathbf{X})\mathcal{S}(Y|\mathbf{X}) \neq \mathcal{T}_{\mathbf{X}}(\mathbf{X})\mathcal{T}(Y|\mathbf{X})$ where the condition $\mathcal{S}(Y|\mathbf{X}) = \mathcal{T}(Y|\mathbf{X})$ while $\mathcal{S}_{\mathbf{X}} = \mathcal{T}_{\mathbf{X}}$ is generally called covariate shift assumption.

Similarity of the (unlabeled) marginal distributions. Ben-David *et al.* [BEN 10b] considered the $\mathcal{H}\Delta\mathcal{H}$-distance between $\mathcal{S}_{\mathbf{X}}$ and $\mathcal{T}_{\mathbf{X}}$ in order to access the impossibility of domain adaptation and assumed that it remains low between the two domains. This is the most straightforward assumption that directly follows from all proposed generalization bounds for domain adaptation. We refer the reader to Chapter 3 for the details.

Ideal joint error. Finally, the last important assumption is the one stating that there should exist a low-error hypothesis for both domains. As explained in Chapter 3, this error can be defined as a so-called $\lambda_{\mathcal{H}}$ term as follows:

$$\lambda_{\mathcal{H}} = \min_{h \in \mathcal{H}} \mathrm{R}_{\mathcal{S}}(h) + \mathrm{R}_{\mathcal{T}}(h).$$

These three assumptions are at the heart of the impossibility theorems where they are analyzed in a pair-wise fashion: both low joint error and small divergence between the marginal distributions assumptions are separately combined with the covariate shift assumption and shown to lead to very low learnability in both cases. More formally, this latter statement can be presented in the form of the following theorems.

4.2. Impossibility theorems

In what follows, we present a series of so-called impossibility results related to the domain adaptation theory. These results are then illustrated based on some concrete examples that highlight the pitfalls of domain adaptation algorithms.

4.2.1. *Theoretical results*

The first theorem shows that some of the intuitive assumptions presented above do not suffice to guarantee the success of domain adaptation. More precisely, among the three assumptions that have been quickly discussed – covariate shift, small $\mathcal{H}\Delta\mathcal{H}$-distance between the unlabeled distributions and the existence of hypotheses that achieve a low error on both the source and target domains (small $\lambda_{\mathcal{H}}$) – the last two are both necessary (and, as we know from previous results, are also sufficient).

THEOREM 4.1 (Necessity of small $\mathcal{H}\Delta\mathcal{H}$-distance [BEN 10b]).– *Let* $\mathbf{X}$ *be some domain set and* $\mathcal{H}$ *a class of functions over* $\mathbf{X}$*. Assume that, for some* $\mathcal{A} \subseteq \mathbf{X}$*, we have that* $\{h^{-1}(1) \cap \mathcal{A} : h \in \mathcal{H}\}$ *contains more than two sets and is linearly ordered by inclusion. Then, the conditions covariate shift plus small* $\lambda_{\mathcal{H}}$ *do not suffice for domain adaptation. In particular, for every* $\epsilon > 0$ *there exists probability distributions* $\mathcal{S}$ *over* $\mathbf{X} \times \{0,1\}$*, and* $\mathcal{T}_{\mathbf{X}}$ *over* $\mathbf{X}$ *such that for every domain adaptation learner* $\mathcal{A}$*, all integers* $m > 0$*,* $n > 0$*, there exists a labeling function* $f : \mathbf{X} \to \{0,1\}$ *such that*

1) $\lambda_{\mathcal{H}} \leq \epsilon$ *is small;*

2) $\mathcal{S}$ *and* $\mathcal{T}_f$ *satisfy the covariate shift assumption;*

3) $\underset{\substack{S\sim(\mathcal{S})^m \\ T_u\sim(\mathcal{T}_{\mathbf{X}})^n}}{\mathbf{Pr}} \left[\mathrm{R}_{\mathcal{T}_f}(\mathcal{A}(S,T_u)) \geq \frac{1}{2}\right] \geq \frac{1}{2}$,

where the distribution $\mathcal{T}_f$ *over* $\mathbf{X} \times \{0,1\}$ *is defined as* $\mathcal{T}_f\{1|\mathbf{x} \in \mathbf{X}\} = f(\mathbf{x})$.

The proof of this theorem can be found in Appendix 2. This result highlights the importance of the need for a small divergence between the marginal distributions of the domains, as even if the covariate shift assumption is satisfied and $\lambda_{\mathcal{H}}$ is small, the error of the classifier returned by a domain adaptation learner can be larger than $\frac{1}{2}$ with probability exceeding this same value. In order to complete this section, we now proceed to the symmetric result that shows the necessity of small joint error between two domains expressed by the $\lambda_{\mathcal{H}}$ term. As mentioned before, this term has a very strong intuition and is expected to reflect the *a priori* capacity of adaptation. The following theorem highlights its importance explicitly.

THEOREM 4.2 (Necessity of small $\lambda_{\mathcal{H}}$ [BEN 10b]).– *Let* $\mathbf{X}$ *be some domain set, and* $\mathcal{H}$ *a class of functions over* $\mathbf{X}$ *whose VC dimension is much smaller than* $|\mathbf{X}|$ *(for instance, any* $\mathcal{H}$ *with a finite VC dimension over an infinite* $\mathbf{X}$*). Then, the conditions covariate shift plus small* $\mathcal{H}\Delta\mathcal{H}$*-divergence do not suffice for domain adaptation. In particular, for every* $\epsilon > 0$ *there exists probability distributions* $\mathcal{S}$ *over* $\mathbf{X}\times\{0,1\}$*, and* $\mathcal{T}_{\mathbf{X}}$ *over* $\mathbf{X}$ *such that for every domain adaptation learner* $\mathcal{A}$*, all integers* $m, n > 0$*, there exists a labeling function* $f : \mathbf{X} \to \{0,1\}$ *such that*

1) $d_{\mathcal{H}\Delta\mathcal{H}}(\mathcal{T}_{\mathbf{X}}, \mathcal{S}_{\mathbf{X}}) \leq \epsilon$ *is small;*

2) the covariate shift assumption holds;

3) $\Pr_{\substack{S\sim\mathcal{S}^m \\ T_u\sim(\mathcal{T}_{\mathbf{X}})^n}} \left[\mathrm{R}_{\mathcal{T}_f}(\mathcal{A}(S, T_u)) \geq \frac{1}{2}\right] \geq \frac{1}{2}.$

The proof of this theorem can be found in Appendix 2. Once again, this theorem shows that small divergence combined with a satisfied covariate shift assumption may lead to an error of the hypothesis returned by a domain adaptation learner that exceeds $\frac{1}{2}$ with high probability. Consequently, the main conclusion of these two theorems can be summarized as follows: among the studied assumptions, neither the assumption combination (1) and (3) nor (2) and (3) suffices for successful domain adaptation in the case of unsupervised domain adaptation. Another important conclusion that should be underlined here is that all generalization bounds for domain adaptation with distance term and joint error term introduced throughout this book indeed imply learnability even with the most straightforward learning algorithm. On the other hand, the covariate shift assumption is rather unnecessary: it cannot replace any of the other assumptions, and it becomes redundant when the other two assumptions hold. This study, however, needs a further investigation, as in the case of semi-supervised domain adaptation the situation can be drastically different.

4.2.2. *Illustrative examples*

Now, as the main impossibility theorems are stated, it can be useful to give an illustrative example of situations where different assumptions and different learning strategies may fail or succeed. To this end, Ben-David *et al.* [BEN 10b] considered several examples that clearly show the inadequacy of the covariate shift assumption explained above as well as the limits of the reweighting scheme.

In what follows, the hypothesis class considered is restricted to the space of threshold functions on $[0,1]$ where a threshold function $h_t(\mathbf{x})$ is defined for any $t \in [0,1]$ as $h_t(\mathbf{x}) = 1$ if $\mathbf{x} < t$ and 0 otherwise. In this case, the set $\mathcal{H}\Delta\mathcal{H}$ becomes the class of half-open intervals.

Inadequacy of the covariate shift. Let us consider the following construction: for some small fixed $\xi \in \{0; 1\}$, let $\mathcal{T}$ be a uniform distribution over $\{2k\xi : k \in \mathbb{N}, 2k\xi \leq 1\} \times \{1\}$ and let the source distribution $\mathcal{S}$ be the uniform distribution over $\{(2k+1)\xi : k \in \mathbb{N}, (2k+1)\xi \leq 1\} \times \{0\}$. The illustration of these distributions is given in Figure 4.1.

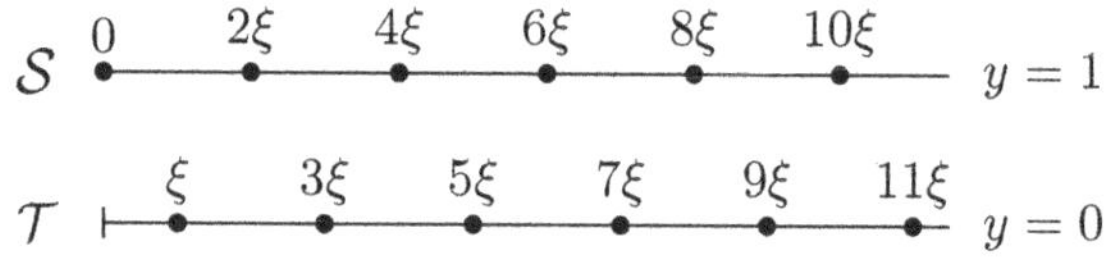

$\mathcal{S}$ 0 2ξ 4ξ 6ξ 8ξ 10ξ $y = 1$

$\mathcal{T}$ ξ 3ξ 5ξ 7ξ 9ξ 11ξ $y = 0$

Figure 4.1. *The scheme illustrates the considered source and target distributions satisfying the covariate shift assumption with* $\xi = \frac{2}{23}$

For this construction, the following holds.

1) The covariate shift assumption holds for $\mathcal{T}$ and $\mathcal{S}$.

2) The distance $d_{\mathcal{H}\Delta\mathcal{H}}(\mathcal{S}, \mathcal{T}) = \xi$ and thus can be arbitrary small.

3) The errors $\mathrm{R}_{\mathcal{S}}(\mathcal{H})$ and $\mathrm{R}_{\mathcal{T}}(\mathcal{H})$ are zero.

4) $\lambda_{\mathcal{H}}(\mathcal{S}, \mathcal{T}) = 1 - \xi$ and $\mathrm{R}_{\mathcal{T}}(h^*_{\mathcal{S}}) \geq 1 - \xi$ are large.

Obviously, from this example one can instantly see that the covariate shift assumption, even combined with a small $\mathcal{H}\Delta\mathcal{H}$-divergence between domains, still results in a large joint error and consequently in a complete failure of the best source classifier when applied to the target distribution.

Reweighting method. A reweighting method in domain adaptation consists of determining a vector of weights $\mathbf{w} = \{w_1, w_2, \ldots, w_m\}$ that are used to reweight the unlabeled source sample S_u generated by $\mathcal{S}_{\mathbf{X}}$ in order to built a new distribution $\mathcal{T}_{\mathbf{w}}^{S_u}$ such that $d_{\mathcal{H}\Delta\mathcal{H}}(\mathcal{T}_{\mathbf{w}}^{S_u}, \mathcal{T}_{\mathbf{X}})$ is as small as possible. In what follows, we denote this reweighted distribution $\mathcal{T}^S$. This new sample is then fed to any available supervised learning algorithm at hand in order to produce a classifier that is expected to have a good performance when applied subsequently in the target domain. As this method plays a very important role in the domain adaptation, the authors also gave two intrinsically close examples that show both its success and failure under the standard domain adaptation assumptions.

We first consider the following scheme: for some small $\epsilon \in \left(0, \frac{1}{4}\right)$, we assume that the covariate shift assumption holds, i.e. for any $\mathbf{x} \in \mathbf{X}$, $\mathcal{T}(y = 1|\mathbf{x}) = \mathcal{S}(y = 1|\mathbf{x}) = f(\mathbf{x})$. We define $f : \mathbf{X} \to [0, 1]$ as follows: for $\mathbf{x} \in [1 - 3\epsilon, 1 - \epsilon]$ we set $f(\mathbf{x}) = 0$ and otherwise we set $f(\mathbf{x}) = 1$. In order to define $\mathcal{S}$ and $\mathcal{T}$, we have only to specify their marginals $\mathcal{S}_{\mathbf{X}}$ and $\mathcal{T}_{\mathbf{X}}$. To this end, we let $\mathcal{S}_{\mathbf{X}}$ be the uniform distribution over $[0, 1]$ and we let $\mathcal{T}_{\mathbf{X}}$ to be the uniform distribution over $[1 - \epsilon, 1]$. This particular setting is depicted in Figure 4.2.

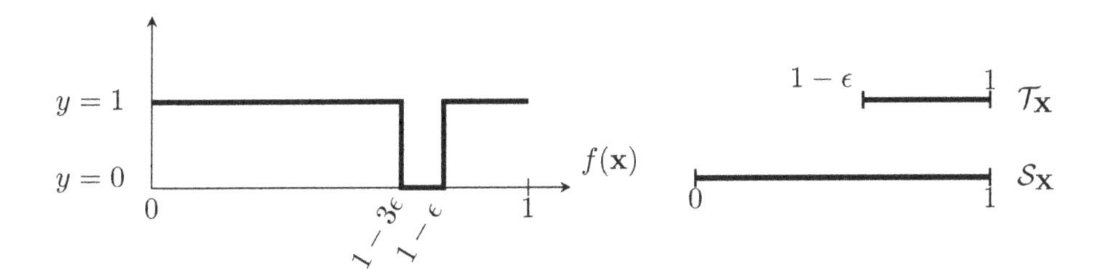

Figure 4.2. *Illustration for the reweighting scenario. Source and target distribution satisfy the covariate shift assumption where f is their common conditional distribution. The marginal $\mathcal{S}_{\mathbf{X}}$ is the uniform distribution over $[0,1]$ and the marginal $\mathcal{T}_{\mathbf{X}}$ is the uniform distribution over $[1-\epsilon,1]$*

The following observations follow from this construction.

1) For the given construction, the best joint hypothesis that defines $\lambda_{\mathcal{H}}$ is given by the function $h_{t=1}$. This function commits 0 errors on target distribution and 2ϵ errors on source distribution, thus giving $\lambda_{\mathcal{H}}(\mathcal{S},\mathcal{T})$ equal to 2ϵ.

2) From the definition of $\mathcal{H}\Delta\mathcal{H}$-divergence, we get that $d_{\mathcal{H}\Delta\mathcal{H}}(\mathcal{S}_{\mathbf{X}},\mathcal{T}_{\mathbf{X}}) = 1-\epsilon$.

3) $\mathrm{R}_{\mathcal{T}}(h^*_{\mathcal{S}}) = 1$, $\mathrm{R}_{\mathcal{T}}(\mathcal{H}) = 0$ and $\mathrm{R}_{\mathcal{S}}(\mathcal{H}) = \epsilon$ achieved by threshold functions $h_{t=1-3\epsilon}$, $h_{t=1}$ and $h_{t=1-3\epsilon}$, respectively.

On the other hand, one can find a reweighting distribution that will produce a sample such that $\mathrm{R}_{\mathcal{T}}(h^*_{\mathcal{T}S}) \to 0$ in probability when m and n tend toward infinity and $h^*_{\mathcal{T}S} = \underset{h\in\mathcal{H}}{\operatorname{argmin}}\ \mathrm{R}_{\mathcal{T}}(h_{\mathcal{T}S})$. This happens along with the probability of the source error tending to 1 when m grows to infinity. This example is a clear illustration when a simple reweighting scheme can be efficient for adaptation. This is not the case however when we consider a different labeling of the target data points. Let us now assume that the source distribution remains the same while for the target distribution $f(\mathbf{x}) = 1$ for any $\mathbf{x} \in \mathbf{X}$. This slight change gives the following results:

1) $\lambda_{\mathcal{H}}(\mathcal{S},\mathcal{T}) = \epsilon$,

2) $d_{\mathcal{H}\Delta\mathcal{H}}(\mathcal{S}_{\mathbf{X}},\mathcal{T}_{\mathbf{X}}) = 1-\epsilon$,

3) $\mathrm{R}_{\mathcal{T}}(h^*_{\mathcal{S}}) = 0$, $\mathrm{R}_{\mathcal{T}}(\mathcal{H}) = 0$ and $\mathrm{R}_{\mathcal{S}}(\mathcal{H}) = \epsilon$.

We can observe that now the $\lambda_{\mathcal{H}}$ term has become smaller and the best source hypothesis achieves a 0 error on the target distribution. However, the result that we obtain with the reweighted method is completely different: it is not hard to see that $\mathrm{R}_{\mathcal{T}}(h^*_{\mathcal{T}S}) \to 1$ in probability when m and n tend toward infinity, while the error of $h^*_{\mathcal{S}}$ will tend to zero.

We conclude by saying that the bound from [BEN 10a] recalled in Chapter 3 implies that $\mathrm{R}_{\mathcal{T}}(h_{\mathcal{S}}^*)$ is bounded by $\mathrm{R}_{\mathcal{T}}(\mathcal{H}) + \lambda_{\mathcal{H}}(\mathcal{S}, \mathcal{T}) + d_{\mathcal{H}\Delta\mathcal{H}}(\mathcal{S}_{\mathbf{X}}, \mathcal{T}_{\mathbf{X}})$ and thus one could have hoped that, by reweighting the sample S to reflect the distribution $\mathcal{T}_{\mathbf{X}}$, the term $d_{\mathcal{H}\Delta\mathcal{H}}(\mathcal{S}_{\mathbf{X}}, \mathcal{T}_{\mathbf{X}})$ in that bound would be diminished. The last example, however, shows that this may not be the case as $\mathrm{R}_{\mathcal{T}_{\mathbf{w}}^{L_{\mathbf{X}}}}$ may be as bad as that bound allows.

4.3. Hardness results

In this section, we present several results that assess the hardness of the domain adaptation problem presented by Ben-David and Urner [BEN 12c]. We first introduce a slightly modified definition of learnability in the context of domain adaptation that is used later.

DEFINITION 4.3.– *Let $\mathcal{W}$ be a class of triples $(\mathcal{S}_{\mathbf{X}}, \mathcal{T}_{\mathbf{X}}, f)$ of source and target distributions over some domain $\mathbf{X}$ and a labeling function, and let $\mathcal{A}$ be a domain adaptation learner. We say that $\mathcal{A}$ $(\varepsilon, \delta, m, n)$-solves domain adaptation for the class $\mathcal{W}$ if, for all triples $(\mathcal{S}_{\mathbf{X}}, \mathcal{T}_{\mathbf{X}}, f) \in \mathcal{W}$, when given access to a sample S_u of size m, generated i.i.d. by $\mathcal{S}_{\mathbf{X}}$ and labeled by f, and an unlabeled sample T_u of size n, generated i.i.d. by $\mathcal{T}_{\mathbf{X}}$, with probability at least $1 - \delta$ (over the choice of the samples S_u and T_u), $\mathcal{A}$ outputs a function h with $\mathrm{R}_{\mathcal{T}}(h) \leq \varepsilon$.*

Compared to definition 4.2, this definition is stated in terms of the probabilistic bound on the target error, while the former bounds the error of the classifier produced using a given domain adaptation learner by the error of the best classifier plus ϵ. Obviously, if we assume that our hypothesis class is rich enough, these two definitions are somehow equivalent as we can always find a zero-error hypothesis. We will further need the following definition which expresses the capacity of a hypothesis class to produce a zero-error classifier with margin γ.

DEFINITION 4.4.– *Let $\mathbf{X} \subseteq \mathbb{R}^d$, $\mathcal{D}_{\mathbf{X}}$be a distribution over $\mathbf{X}$ and $h : \mathbf{X} \to \{0, 1\}$ be a classifier and $B_\gamma(\mathbf{x})$ be the ball of radius γ around some domain point $\mathbf{x}$. We say that h is a γ-margin classifier with respect to $\mathcal{D}_{\mathbf{X}}$ if for all $\mathbf{x} \in \mathbf{X}$ whenever $\mathcal{D}_{\mathbf{X}}(B_\gamma(\mathbf{x})) > 0$, $h(y) = h(z)$ holds for all $y, z \in B_\gamma(\mathbf{x})$.*

Ben-David *et al.* [BEN 12c] also note that h being a γ-margin classifier with respect to $\mathcal{D}_{\mathbf{X}}$ is equivalent to h satisfying the Lipschitz-property with Lipschitz constant $\frac{1}{2\gamma}$ on the support of $\mathcal{D}_{\mathbf{X}}$. Thus, we may refer to this assumption as the Lipschitzness assumption. For the sake of completeness, we present the original definition of the probabilistic Lipschitzness below.

DEFINITION 4.5.– *Let $\phi : \mathbb{R} \to [0,1]$. We say that $f : \mathbf{X} \to \mathbb{R}$ is ϕ-Lipschitz with respect to a distribution $\mathcal{D}_\mathbf{X}$ over $\mathbf{X}$ if, for all $\lambda > 0$, we have*

$$\Pr_{\mathbf{x} \sim \mathcal{D}_\mathbf{X}} \left[\exists \mathbf{x}' : |f(\mathbf{x}) - f(\mathbf{x}')| > \lambda \mu(\mathbf{x}, \mathbf{x}') \right] \leq \phi(\lambda),$$

where $\mu : \mathbf{X} \times \mathbf{X} \to \mathbb{R}_+$ is some metric over $\mathbf{X}$.

The first hardness results for domain adaptation can be stated as follows.

THEOREM 4.3 ([BEN 12c]).– *For every finite domain $\mathbf{X}$, for every ε and δ with $\varepsilon + \delta < \frac{1}{2}$, no algorithm can $(\varepsilon, \delta, s, t)$-solve the domain adaptation problem for the class $\mathcal{W}$ of triples $(\mathcal{S}_\mathbf{X}, \mathcal{T}_\mathbf{X}, f)$ with $C_\mathcal{B}(\mathcal{S}_\mathbf{X}, \mathcal{T}_\mathbf{X}) \geq \frac{1}{2}$, $d_{\mathcal{H}\Delta\mathcal{H}}(\mathcal{S}_\mathbf{X}, \mathcal{T}_\mathbf{X}) = 0$ and $\mathrm{R}_\mathcal{T}(\mathcal{H}) = 0$ if*

$$s + t < \sqrt{(1 - 2(\varepsilon + \delta))|\mathbf{X}|},$$

where $\mathcal{H}$ is the hypothesis class that contains only the all-1 and the all-0 labeling functions, $\mathrm{R}_\mathcal{T}(\mathcal{H}) = \min_{h \in \mathcal{H}} \mathrm{R}_\mathcal{T}(h, f)$ and

$$C_\mathcal{B}(\mathcal{S}_\mathbf{X}, \mathcal{T}_\mathbf{X}) = \inf_{\substack{b \in \mathcal{B} \\ \mathcal{T}_\mathbf{X}(b) \neq 0}} \frac{\mathcal{S}_\mathbf{X}(b)}{\mathcal{T}_\mathbf{X}(b)}$$

is a weight ratio [BEN 12c] of source and target domains, with respect to a collection of input space subsets $\mathcal{B} \subseteq 2^\mathbf{X}$.

The proof of this theorem can be found in Appendix 2. This result is interesting in many ways. First, the assumptions made in the theorem are extremely simplified, which means that the *a priori* knowledge about the target task is so strong that a zero-error classifier for the given hypothesis class can be obtained using only one labeled target instance. Second, we may also note that the considered setting is extremely favorable for adaptation as the marginal distributions of source and target domains are close both in terms of the $\mathcal{H}\Delta\mathcal{H}$-divergence and the weight ratio $C_\mathcal{B}(\mathcal{S}_\mathbf{X}, \mathcal{T}_\mathbf{X})$, where the assumption regarding the latter roughly means that the probability of encountering a source point is at least half of the probability of finding it in the target domain. These assumptions further spur the following surprising conclusions.

– The sample complexity of domain adaptation cannot be bounded only in terms of the VC dimension of the class that can produce a hypothesis achieving a zero error on it. This statement agrees well with the previous results, showing the necessity of the existence of a good hypothesis for both domains.

– Some data drawn from target distribution should be available in order to obtain a bound with an exclusive dependency on the VC dimension of the hypothesis class.

– This result implies that the sample sizes needed to obtain useful approximations of weight ratio is prohibitively high.

We now proceed to another result provided by Ben-David and Urner that shows that the same lower bound can be obtained using the Lipschitzness assumption imposed on the labeling function f.

THEOREM 4.4 ([BEN 12c]).– *Let* $\mathbf{X} = [0,1]^d$, $\varepsilon > 0$ *and* $\delta > 0$ *be such that* $\varepsilon + \delta < \frac{1}{2}$, $\lambda > 1$ *and* $\mathcal{W}_\lambda$ *be the set of triples* $(\mathcal{S}_\mathbf{X}, \mathcal{T}_\mathbf{X}, f)$ *of distributions over* $\mathbf{X}$ *with* $\mathrm{R}_\mathcal{T}(\mathcal{H}) = 0$, $C_\mathcal{B}(\mathcal{S}_\mathbf{X}, \mathcal{T}_\mathbf{X}) \geq \frac{1}{2}$, $d_{\mathcal{H}\Delta\mathcal{H}}(\mathcal{S}_\mathbf{X}, \mathcal{T}_\mathbf{X}) = 0$ *and* λ*-Lipschitz labeling functions* f. *Then no domain adaptation-learner can* $(\varepsilon, \delta, s, t)$*-solve the domain adaptation problem for the class* $\mathcal{W}_\lambda$ *unless*

$$s + t \geq \sqrt{(\lambda+1)^d(1 - 2(\varepsilon + \delta))}.$$

The proof of this theorem can be found in Appendix 2.

4.4. Usefulness of unlabeled data

So far we have presented theorems that show what conditions probably lead to the failure of domain adaptation. These results showed that even in some extremely simple settings successful adaptation may require an abundant number of labeled source data or at least a reasonable amount of labeled target data. In spite of this, a natural question that one may ask is to what extent a target domain's unlabeled data can help to adapt. Before answering this question, the authors consider a particular adaptation algorithm $\mathcal{A}$ that can be summarized as follows.

Algorithm 1 — Input: An i.i.d. sample $S_u \sim (\mathcal{S}_\mathbf{X})^m$ labeled by f, an unlabeled i.i.d. sample $T_u \sim (\mathcal{T}_\mathbf{X})^n$, and a margin parameter γ.

Step 1. Partition the domain $[0,1]^d$ into a collection $\mathcal{B}$ of boxes (axis-aligned rectangles) with sidelength $\gamma/\sqrt{d}$.
Step 2. Obtain sample S' by removing every point in S_u, which is sitting in a box that is not hit by T_u.
Step 3. Output an optimal risk-minimizing classifier from $\mathcal{H}$ for the sample S'.

The following theorems give a comprehensive answer to this question by providing the lower bounds for both the required size of the source labeled sample and the unlabeled target required for algorithm $\mathcal{A}$ to learn well.

THEOREM 4.5 ([BEN 12c]).– *Let* $\mathbf{X} = [0,1]^d$, $\gamma > 0$ *be a margin parameter,* $\mathcal{H}$ *be a hypothesis class of finite VC dimension and* $\mathcal{W}$ *be the set of triples* $(\mathcal{S}_\mathbf{X}, \mathcal{T}_\mathbf{X}, f)$ *of the source distribution, target distribution and labeling function with*

1) $C_{\mathcal{I}}(\mathcal{S}_{\mathbf{X}}, \mathcal{T}_{\mathbf{X}}) \geq \frac{1}{2}$ *for the class* $\mathcal{I} = (\mathcal{H}\Delta\mathcal{H}) \cap \mathcal{B}$*, where* $\mathcal{B}$ *is a partition of* $[0,1]^d$ *into boxes of sidelength* $\frac{\gamma}{\sqrt{d}}$*;*

2) $\mathcal{H}$ *contains a hypothesis that has* γ*-margin on* $\mathcal{T}$*;*

3) the labeling function f *is a* γ*-margin classifier with respect to* $\mathcal{T}$*.*

Then there is a constant $c > 1$ *such that, for all* $\varepsilon > 0$*,* $\delta > 0$ *and for all* $(\mathcal{S}_{\mathbf{X}}, \mathcal{T}_{\mathbf{X}}, f) \in \mathcal{W}$*, when given an i.i.d. sample* S_u *from* $\mathcal{S}_{\mathbf{X}}$*, labeled by* f *of size*

$$|S_u| \geq c\left[\frac{VC(\mathcal{H}) + \log\frac{1}{\delta}}{C_{\mathcal{I}}(\mathcal{S}_{\mathbf{X}}, \mathcal{T}_{\mathbf{X}})(1-\varepsilon)\varepsilon}\log\left(\frac{VC(\mathcal{H})}{C_{\mathcal{I}}(\mathcal{S}_{\mathbf{X}}, \mathcal{T}_{\mathbf{X}})(1-\varepsilon)\varepsilon}\right)\right],$$

and an i.i.d. sample T_u *from* $\mathcal{T}_{\mathbf{X}}$ *of size*

$$|T_u| \geq \frac{1}{\epsilon}\left(2\left[\frac{\sqrt{d}}{\gamma}\right]^d \ln\left(3\left[\frac{\sqrt{d}}{\gamma}\right]^d \delta\right)\right),$$

$\mathcal{A}$ *outputs a classifier* h *with* $\mathrm{R}_{\mathcal{T}}(h,f) \leq \epsilon$ *with probability at least* $1-\delta$*.*

The proof of this theorem can be found in Appendix 2. It is worth noticing that these bounds follow the standard bounds from the statistical learning theory where the size of the learning sample required for successful learning is given as a function of the VC dimension of the hypothesis class. In domain adaptation, this dependency is further extended to the weight ratio and the accuracy parameters of the learnability model. Moreover, we observe that this theorem considers the input space that may contain an infinite number of points. This assumption can lead to a vacuous bound as in reality the input space often presents a finite domain and the dependency of the sample size should be given in its terms. The following theorem covers this case.

THEOREM 4.6.– *Let* $\mathbf{X}$ *be some finite domain,* $\mathcal{H}$ *be a hypothesis class of finite VC dimension and* $\mathcal{W} = \{(\mathcal{S}_{\mathbf{X}}, \mathcal{T}_{\mathbf{X}}, f) | \mathrm{R}_{\mathcal{T}}(\mathcal{H}) = 0, C(\mathcal{S}_{\mathbf{X}}, \mathcal{T}_{\mathbf{X}}) \geq 0\}$ *be a class of pairs of source and target distributions with bounded weight ratio where the* $\mathcal{H}$ *contains a zero-error hypothesis on* $\mathcal{T}$*. Then there is a constant* $c > 1$ *such that, for all* $\varepsilon > 0$*,* $\delta > 0$*, and all* $(\mathcal{S}_{\mathbf{X}}, \mathcal{T}_{\mathbf{X}}, f) \in \mathcal{W}$*, when given an i.i.d. sample* S_u *from* $\mathcal{S}_{\mathbf{X}}$*, labeled by* f *of size*

$$|S_u| \geq c\left[\frac{VC(\mathcal{H}) + \log\frac{1}{\delta}}{C(\mathcal{S}_{\mathbf{X}}, \mathcal{T}_{\mathbf{X}})(1-\varepsilon)\varepsilon}\log\left(\frac{VC(\mathcal{H})}{C(\mathcal{S}_{\mathbf{X}}, \mathcal{T}_{\mathbf{X}})(1-\varepsilon)\varepsilon}\right)\right],$$

and an i.i.d. sample T_u *from* $\mathcal{T}_{\mathbf{X}}$ *of size*

$$|T_u| \geq \frac{1}{\epsilon}\left(\frac{2|\mathbf{X}|\ln 3|\mathbf{X}|}{\delta}\right),$$

algorithm $\mathcal{A}$ *outputs a classifier* h *with* $\mathrm{R}_{\mathcal{T}}(h,f) \leq \epsilon$ *with probability at least* $1-\delta$*.*

To conclude, we note that both hardness results that state under which conditions domain adaptation fails, as well as the results of the analysis of the samples' sizes required from source and target domains for adaptation to succeed, fall into the category of the so-called impossibility theorems. They essentially draw the limits of the domain adaptation problem under various common assumptions and provide insights into the hardness of solving it.

To proceed, we now turn our attention to a series of impossibility results presented by Ben-David *et al.* [BEN 12b] that investigate the existence of a learning method that is capable of efficiently learning a good hypothesis for a target task provided that the target sample from its corresponding probability distribution is replaced by a (possibly larger) generated sample from a different probability distribution. The efficiency of such a learning method requires it not to worsen the generalization guarantee of the learned classifier in the target domain. As an example of the considered classifier, we can take a popular nearest-neighbor classifier $h_{\text{NN}}(\mathbf{x})$ that given a metric μ defined over the input space $\mathbf{X}$ assigns a label to a point $\mathbf{x}$ as $h_{\text{NN}}(\mathbf{x}) = y(N_S(\mathbf{x}))$, with $N_S(\mathbf{x}) = \text{argmin}_{\mathbf{z} \in S}\, \mu(\mathbf{x}, \mathbf{z})$ being the nearest neighbor of $\mathbf{x}$ in the labeled source sample S, and $y(N_S(\mathbf{x}))$ being the label of this nearest neighbor. The obtained theorems are proved under the covariate shift condition and assume a bound weight ratio between the two domains as explained before. We now present the first theorem proved for this case below.

THEOREM 4.7 ([BEN 12b]).– *Let domain $\mathbf{X} = [0,1]^d$ and for some $C > 0$, let $\mathcal{W}$ be a class of pairs of source and target distributions $\{(\mathcal{S}, \mathcal{T}) | C_{\mathcal{B}}(\mathcal{S}_{\mathbf{X}}, \mathcal{T}_{\mathbf{X}}) \geq C\}$ with bounded weight ratio and their common labeling function $f : \mathbf{X} \to [0,1]$, satisfying the ϕ-probabilistic-Lipschitz property with respect to the target distribution, for some function ϕ. Then, for all λ,*

$$\mathop{\mathbf{E}}_{S \sim \mathcal{S}^m} [\mathrm{R}_{\mathcal{T}}(h_{NN})] \leq 2\mathrm{R}^*_{\mathcal{T}}(\mathcal{H}) + \phi(\lambda) + 4\lambda \frac{\sqrt{d}}{C} m^{-\frac{1}{d-1}} .$$

The proof of this theorem can be found in Appendix 2. This theorem suggests that under several assumptions (i.e. covariate shift and bounded weight ratio) the expected target error of an NN classifier learned on a sample drawn from the source distribution is bounded by twice the optimal risk over the whole considered hypothesis space plus several constants related to the nature of the labeling function and the dimension of the input space. Regarding these latter, one may note that if the labeling function is λ-Lipschitz in the standard sense of Lipschitzness and the labels are deterministic, then we have $\mathrm{R}^*_{\mathcal{T}}(\mathcal{H}) = 0$ and $\phi(a) = 0$ for all $a \geq \lambda$. Applying Markov's inequality then yields the following corollary on sample size bound which further strengthens the previous result.

COROLLARY 4.1.– *Let domain* $\mathbf{X} = [0,1]^d$ *and for some* $C > 0$, *let* $\mathcal{W}$ *be a class of pairs of source and target distributions* $\{(\mathcal{S},\mathcal{T})|C_{\mathcal{B}}(\mathcal{S}_{\mathbf{X}},\mathcal{T}_{\mathbf{X}}) \geq C\}$ *with bounded weight ratio and their common labeling function* $f : \mathbf{X} \to [0,1]$ *satisfying the* ϕ*-probabilistic-Lipschitz property with respect to the target distribution, for some function* ϕ. *Then, for all* $\varepsilon > 0$, $\delta > 0$, $m \geq \left(\frac{4\lambda\sqrt{d}}{C\varepsilon\delta}\right)^{d+1}$ *the nearest neighbor algorithm applied to a sample of size* m, *has, with probability of at least* $1-\delta$, *error of at most* ε *w.r.t. the target distribution for any pair* $(\mathcal{S},\mathcal{T}) \in \mathcal{W}$.

4.5. Proper domain adaptation

In this part, we present other impossibility results established in [BEN 12b] for the case where the output of the given DA algorithm should be a hypothesis belonging to some predefined hypothesis class. This particular constraint easily justifies itself in practice where one may want to find a hypothesis as quickly as possible from a predefined set of hypotheses at the expense of having a higher error rate. In order to proceed, we first define the proper DA setting in the following definition.

DEFINITION 4.6.– *If* $\mathcal{H}$ *is a class of hypotheses, we say that* $\mathcal{A}$ $(c,\varepsilon,\delta,m,n)$*-solves proper DA for the class* $\mathcal{W}$ *relative to* $\mathcal{H}$ *if, for all pairs* $(\mathcal{S},\mathcal{T}) \in \mathcal{W}$, *when given access to a labeled sample* S *of size* m, *generated i.i.d. by* $\mathcal{S}$, *and an unlabeled sample* T_u *of size* n, *generated i.i.d. by* $\mathcal{T}_{\mathbf{X}}$, *with probability at least* $1-\delta$ *(over the choice of the samples* S *and* T_u*),* $\mathcal{A}$ *outputs an element* h *of* $\mathcal{H}$ *with*

$$\mathrm{R}_{\mathcal{T}}(h) \leq c\mathrm{R}_{\mathcal{T}}(\mathcal{H}) + \varepsilon.$$

In other words, this definition says that proper solving of the DA problem is achieved when the error of the returned hypothesis on the target distribution is bounded by the c times the error of the best hypothesis on the target distribution plus a constant ε. Obviously, efficient solving of the proper DA is characterized by small δ, ϵ and c close to 1.

The following result was obtained in [BEN 12c] in this setting.

THEOREM 4.8 ([BEN 12b]).– *Let domain* $\mathbf{X} = [0,1]^d$ *for some* d. *Consider the class* $\mathcal{H}$ *of half-spaces as the target class. Let* $\mathbf{x}$ *and* $\mathbf{z}$ *be a pair of antipodal points on the unit sphere and* $\mathcal{W}$ *be a set that contains two pairs* $(\mathcal{S},\mathcal{T})$ *and* $(\mathcal{S}',\mathcal{T}')$ *of distributions where:*

1) both pairs satisfy the covariate shift assumption;

2) $f(\mathbf{x}) = f(\mathbf{z}) = 1$ *and* $f(\bar{0}) = 0$ *for their common labeling function* f*;*

3) $\mathcal{S}_{\mathbf{X}}(\mathbf{x}) = \mathcal{T}_{\mathbf{X}}(\mathbf{z}) = \mathcal{S}_{\mathbf{X}}(\bar{0}) = \frac{1}{3}$*;*

4) $\mathcal{T}_{\mathbf{X}}(\mathbf{x}) = \mathcal{T}_{\mathbf{X}}(\bar{0}) = \frac{1}{2}$ *or* $\mathcal{T}'_{\mathbf{X}}(\mathbf{z}) = \mathcal{T}'_{\mathbf{X}}(\bar{0}) = \frac{1}{2}$.

Then, for any number m, any constant c, no proper DA learning algorithm can $(c, \varepsilon, \delta, m, 0)$-solve the domain adaptation learning task for $\mathcal{W}$ with respect to $\mathcal{H}$, if $\varepsilon < \frac{1}{2}$ and $\delta < \frac{1}{2}$. In other words, every learner that ignores unlabeled target data fails to produce a zero-risk hypothesis with respect to $\mathcal{W}$.

The proof of this theorem can be found in Appendix 2. This theorem shows that having some amount of data generated by the target distribution is crucial for the learning algorithm to estimate whether the support of the target distribution is $\mathbf{x}$ and $\overline{0}$ or $\mathbf{z}$ and $\overline{0}$. Surprisingly, there is no possible way of obtaining this information without having access to a sample drawn from the target distribution event if the point-wise weight ratio is assumed to be as large as $\frac{1}{2}$. Thus, no amount of labeled source data can compensate for having a sample from the target marginal distribution.

Before presenting the last results of this section, we first introduce the definition of agnostic learning.

DEFINITION 4.7 ([BEN 12b]).– *For $\varepsilon > 0, \delta > 0$, $m \in \mathbb{N}$ we say that an algorithm (ε, δ, m) (agnostically) learns a hypothesis class $\mathcal{H}$, if for all distributions $\mathcal{D}$, when given an i.i.d. sample of size at least m, it outputs a classifier of error at most $\mathrm{R}_{\mathcal{D}}(\mathcal{H}) + \varepsilon$ with probability at least $1 - \delta$. If the output of the algorithm is always a member of $\mathcal{H}$, we call it an agnostic proper learner for $\mathcal{H}$.*

Note that this definition is similar to definition 4.6 with the only difference being that it reflects the required property of the learner to produce a hypothesis that belongs to some predefined hypothesis class. This definition leads to the following theorem.

THEOREM 4.9 ([BEN 12b]).– *Let $\mathbf{X}$ be some domain and $\mathcal{W}$ be a class of pairs $(\mathcal{S}, \mathcal{T})$ of distributions over $\mathbf{X} \times \{0,1\}$ with $\mathrm{R}_{\mathcal{T}}(\mathcal{H}) = 0$ such that there is an algorithm $\mathcal{A}$ and functions $m : (0,1)^2 \to \mathbb{N}$, $n : (0,1)^2 \to \mathbb{N}$ such that $\mathcal{A}$ $(0, \varepsilon, \delta, m(\varepsilon, \delta), n(\varepsilon, \delta))$-solves the domain adaptation learning task for $\mathcal{W}$ for all $\varepsilon, \delta > 0$. Let $\mathcal{H}$ be some hypotheses class for which there exists an agnostic proper learner. Then, the $\mathcal{H}$-proper domain adaptation problem can be $((0, \varepsilon, \delta, m(\varepsilon/3, \delta/2), n(\varepsilon/3, \delta/2)) + m'(\varepsilon/3, \delta/2))$-solved with respect to the class $\mathcal{W}$, where m' is the sample complexity function for agnostically learning $\mathcal{H}$.*

The proof of this theorem can be found in Appendix 2. As in the previous case, one can consider the algorithm $\mathcal{A}$ in the statement of this theorem to be nearest neighbor classifier NN($\mathcal{S}$), if the class $\mathcal{W}$ satisfies the conditions from the theorem. To summarize, the presented theorems for the proper DA learning show that with a DA algorithm, which takes into account the unlabeled instances from the target marginal distribution, one may hope to solve the proper DA problem, while in the opposite case it is probably unsolvable.

4.6. Summary

In this chapter, we covered a series of results that establish the necessary conditions required to make a domain adaptation problem solvable. As it was shown, the necessary conditions for the success of domain adaptation may take different forms and depend on the value of certain terms presented in the generalization bound and on the size of the available source and target learning samples. The take-away messages of this chapter can be summarized as follows.

1) Solving a domain adaptation problem requires two independent conditions to be fulfilled. First, we have to properly minimize the divergence between the source and target marginal distributions to make it as small as possible. Second, we have to simultaneously ensure that the *a priori* adaptability of the two domains is high (that is reflected by the small ideal joint error term $\lambda_{\mathcal{H}}$).

2) Even under some strong assumptions that make the adaptation problem seemingly easy to solve, we may still need a certain amount of unlabeled source and target data that, in the most general case, can be prohibitively large.

3) A certain amount of labeled source and unlabeled target data can ensure efficient adaptation and produce a hypothesis with a small target error. In both cases, this amount depends on the general characteristics of the adaptation problem given by the weight ratio and the complexity of the hypothesis space represented by its VC dimension.

4) In proper DA, ignoring unlabeled target data leads to provably unsolvable adaptation problems where the domain adaptation learner fails to produce a zero-error hypothesis for the target domain.

All these conclusions provide us with a more general view on the learning properties of the adaptation phenomenon and essentially give a list of conditions that we have to verify in order to make sure that the adaptation problem at hand can be solved efficiently. Apart from that, the established results also provide us with an understanding that some adaptation tasks are harder than others and that this hardness can be quantified by not one but several criteria that take into account both data distribution and the labeling of instances. Finally, they also show that successful adaptation requires a certain amount of data to be available during the adaptation step and that this amount may directly depend on the proximity of the marginal distributions of the two domains. This last feature is quite important as it is added to the dependence on the complexity of the hypothesis class considered previously in the standard supervised learning described in Chapter 1.

5

Generalization Bounds with Integral Probability Metrics

In previous chapters, we presented several seminal results regarding the generalization bounds for domain adaptation and the impossibility theorems for some of them. We have shown that the basic shape of generalization bounds in the context of domain adaptation remains more or less the same and principally differs only in the divergence used to measure the distance between the source and the target marginal distributions. To this end, we showed that the first original bound proposed by Ben-David *et al.* [BEN 07, BEN 10a] is restricted to a particular $0-1$ loss function and a symmetric hypothesis class: a drawback that was tackled by Mansour *et al.* [MAN 09a] with the introduction of the discrepancy distance. Despite the differences between the two, however, they both have several common features, which is an explicit dependence on the considered hypothesis class and the computational issues related to their computation. Indeed, the computation of the $\mathcal{H}\Delta\mathcal{H}$-divergence presents an intractable problem, while the minimization of the discrepancy distance has a prohibitive computational complexity. To this end, a natural quest for other metrics with some attractive properties suitable for quantifying the divergence between the two domains arises. In this chapter, we consider a large family of metrics on the space of probability measure called integral probability metrics (IPMs) that present a well-studied topic in probability theory. We particularly show that depending on the chosen functional class, some instances of IPMs can have interesting properties that are completely different from those exhibited by both $\mathcal{H}\Delta\mathcal{H}$-divergence and discrepancy distance.

5.1. Definition

In order to understand how the IPMs were introduced, we may proceed as follows. Assume that we have two probability distributions $\mathcal{S}_{\mathbf{X}}$ and $\mathcal{T}_{\mathbf{X}}$ defined over some space $\mathbf{X}$. Then, it is obvious that for any (measurable) function f, if $\mathcal{S}_\mathbf{X} = \mathcal{T}_\mathbf{X}$ we immediately obtain:

$$\mathop{\mathbf{E}}_{\mathbf{x}\sim\mathcal{S}_\mathbf{X}}[f(\mathbf{x})] = \mathop{\mathbf{E}}_{\mathbf{x}\sim\mathcal{T}_\mathbf{X}}[f(\mathbf{x})].$$

On the other hand, one may wonder if this statement also holds in the opposite direction: given two probability distributions $\mathcal{S}_\mathbf{X}$ and $\mathcal{T}_\mathbf{X}$ such that $\mathcal{S}_\mathbf{X} \neq \mathcal{T}_\mathbf{X}$, can we find a function f (usually called witness function) satisfying

$$\mathop{\mathbf{E}}_{\mathbf{x}\sim\mathcal{S}_\mathbf{X}}[f(\mathbf{x})] \neq \mathop{\mathbf{E}}_{\mathbf{x}\sim\mathcal{T}_\mathbf{X}}[f(\mathbf{x})]?$$

If this is the case, one can measure the divergence between $\mathcal{S}_\mathbf{X}$ and $\mathcal{T}_\mathbf{X}$ by considering the difference between the expectations calculated with respect to each of the distributions over some function f. The answer to this question can be given by the following lemma.

LEMMA 5.1 ([MÜL 97]).– *Two distributions $\mathcal{S}_\mathbf{X}$ and $\mathcal{T}_\mathbf{X}$ are identical if and only if for every continuous and differentiable function f*

$$\mathop{\mathbf{E}}_{\mathbf{x}\sim\mathcal{S}_\mathbf{X}}[f(\mathbf{x})] = \mathop{\mathbf{E}}_{\mathbf{x}\sim\mathcal{T}_\mathbf{X}}[f(\mathbf{x})].$$

The general definition of IPMs can be thus given as follows.

DEFINITION 5.1 ([ZOL 84]).– *Given two probability measures $\mathcal{S}_\mathbf{X}$ and $\mathcal{T}_\mathbf{X}$ defined on a measurable space $\mathbf{X}$, IPM is defined as*

$$D_\mathcal{F}(\mathcal{S}_\mathbf{X}, \mathcal{T}_\mathbf{X}) = \sup_{f\in\mathcal{F}} \left| \int_\mathbf{X} f d\mathcal{S}_\mathbf{X} - \int_\mathbf{X} f d\mathcal{T}_\mathbf{X} \right|,$$

where $\mathcal{F}$ is a class of real-valued bounded measurable functions on $\mathbf{X}$.

IPMs present a large class of distances defined on the space of probability measures that find their application in many machine learning algorithms. As mentioned by [MÜL 97], the quantity $D_\mathcal{F}(\mathcal{S}_\mathbf{X}, \mathcal{T}_\mathbf{X})$ is a semimetric, and it is a metric if and only if the function class $\mathcal{F}$ separates the set of all signed measures with $\mu(\mathbf{X}) = 0$. It then follows for any non-trivial function class $\mathcal{F}$ that the quantity $D_\mathcal{F}(\mathcal{S}_\mathbf{X}, \mathcal{T}_\mathbf{X})$ is equal to zero if $\mathcal{S}_\mathbf{X}$ and $\mathcal{T}_\mathbf{X}$ are the same.

5.2. General setting

We start this chapter with a general result that introduces IPMs to the DA generalization bounds provided in [ZHA 12]. In this paper, authors consider a general multisource scenario where not one but $K \geq 2$ source domains are available. In order to be consistent with the rest of the book, we present the main result of Zhang *et al.* [ZHA 12] that introduces the IPMs in the context of domain adaptation specified for the case of one source and one target domain below[1].

THEOREM 5.1.– *For a labeling function $f \in \mathcal{G}$, let $\mathcal{F} = \{(\mathbf{x}, y) \rightarrow \ell(f(\mathbf{x}), y)\}$ be a loss function class consisting of the bounded functions with the range $[a, b]$ for a space of labeling functions $\mathcal{G}$. Let $S = \{(\mathbf{x}_1, y_1), \ldots, (\mathbf{x}_m, y_m)\}$ be a labeled sample drawn from $\mathcal{S}$ of size m. Then given any arbitrary $\xi \geq D_{\mathcal{F}}(\mathcal{S}, \mathcal{T})$, for any $m \geq \frac{8(b-a)}{\xi'^2}$ and any $\epsilon > 0$, with probability at least $1 - \epsilon$, the following holds:*

$$\sup_{f \in \mathcal{F}} \left| \mathrm{R}^{\ell}_{\hat{\mathcal{S}}} f - \mathrm{R}^{\ell}_{\mathcal{T}} f \right| \leq D_{\mathcal{F}}(\mathcal{S}, \mathcal{T}) + \left(\frac{\ln \mathcal{N}_1(\xi'/8, \mathcal{F}, 2m) - \ln(\epsilon/8)}{\frac{m}{32(b-a)^2}} \right)^{\frac{1}{2}},$$

where $\xi' = \xi - D_{\mathcal{F}}(\mathcal{S}, \mathcal{T})$.

Here, the quantity $\mathcal{N}_1(\xi, \mathcal{F}, 2m)$ is defined in terms of the uniform entropy number[2] and is given by the following equation:

$$\mathcal{N}_1(\xi, \mathcal{F}, 2m) = \sup_{\{S^{2m}\}} \log N\big(\xi, \mathcal{F}, \ell_1(\{S^{2m}\})\big),$$

where for source sample S and its associated ghost sample $S' = \{(\mathbf{x}'_1, y'_1), \ldots, (\mathbf{x}'_m, y'_m)\}$ drawn from $\mathcal{S}$ the quantity $S^{2m} = \{S, S'\}$ and the metric ℓ_1 is a variation of the ℓ_1 metric defined for some $f \in \mathcal{F}$ based on the following norm:

$$\|f\|_{\ell_1(\{S^{2m}\})} = \frac{1}{m} \sum_{i=1}^{m} \Big(|f(\mathbf{x}_i, y_i)| + |f(\mathbf{x}'_i, y_i)| \Big).$$

One may note several peculiarities related to this result. First, it is different from other generalization bounds provided before as the divergence term here is defined for joint distributions $\mathcal{S}$ and $\mathcal{T}$ and not for marginal distributions $\mathcal{S}_{\mathbf{X}}$ and $\mathcal{T}_{\mathbf{X}}$. Note that, in general, one cannot estimate the joint target distribution $\mathcal{T}$ in the classical scenario of unsupervised domain adaptation as this can be done only when target labels are known, thus making the application of this bound quite uninformative in practice. Second, the proposed bound is very general as it does not explicitly specify the functional class $\mathcal{F}$ considered in the definition of the IPM. On the one hand, this allows us to adjust this bound to any instance of IPMs that can be obtained by choosing the appropriate functional class but, on the other hand, it also requires us to determine the uniform entropy number for it. Finally, the authors establish a link

1 The general formulations of theorems 5.1 and 5.2 for K distinct source domains are given in Appendix 3.

2 We recall that the definition of a covering number was given in definition 1.7.

between the discrepancy distance seen before and the $D_{\mathcal{F}}(\mathcal{S},\mathcal{T})$ that allows us to obtain a bound that has a more "traditional" shape. More precisely, the authors proved that the following inequality holds in the case of one source and one target domain for any ℓ and functional class $\mathcal{F}$:

$$D_{\mathcal{F}}(\mathcal{S},\mathcal{T}) \leq disc_{\ell}(\mathcal{S}_{\mathbf{X}},\mathcal{T}_{\mathbf{X}}) + \sup_{g \subset \mathcal{G}} \left| \mathop{\mathbf{E}}_{\mathbf{x}\sim\mathcal{T}_{\mathbf{X}}} [\ell(g(\mathbf{x}), f_{\mathcal{T}}(\mathbf{x}))] - \mathop{\mathbf{E}}_{\mathbf{x}\sim\mathcal{T}_{\mathbf{X}}} [\ell(g(\mathbf{x}), f_{\mathcal{S}}(\mathbf{x}))] \right|.$$

Note that the second term of the right-hand side is basically a disagreement between the labeling functions $f_{\mathcal{S}}$ and $f_{\mathcal{T}}$ that is equal to zero only when they are equal. Using this inequality, one may show that the proposed theorem can be reduced to have the following shape:

$$\sup_{f\in\mathcal{F}} |\mathrm{R}^{\ell}_{\hat{\mathcal{S}}} f - \mathrm{R}^{\ell}_{\mathcal{T}} f| \leq disc_{\ell}(\mathcal{S}_{\mathbf{X}},\mathcal{T}_{\mathbf{X}}) + \lambda + \left(\frac{\ln \mathcal{N}_1(\xi'/8,\mathcal{F},2m) - \ln(\epsilon/8)}{\frac{m}{32(b-a)^2}} \right)^{\frac{1}{2}}, \qquad [5.1]$$

where $\lambda = \sup_{g\in\mathcal{G}} \left| \mathop{\mathbf{E}}_{\mathbf{x}\sim\mathcal{T}_{\mathbf{X}}} [\ell(g(\mathbf{x}), f_{\mathcal{T}}(\mathbf{x}))] - \mathop{\mathbf{E}}_{\mathbf{x}\sim\mathcal{T}_{\mathbf{X}}} [\ell(g(\mathbf{x}), f_{\mathcal{S}}(\mathbf{x}))] \right|$ and the last term is the complexity term that depends on the covering number of the space $\mathcal{F}$, similar to the bounds based on algorithmic robustness presented in equation [1]. To this end, equation [5.1] now looks pretty much like the generalization bounds from the previous chapters.

In order to show that for a finite complexity term, the difference between the empirical source risk and the target risk never exceeds the divergence between the two domains with the increasing number of available source examples, the authors prove the following theorem.

THEOREM 5.2.– *For a labeling function $f \in \mathcal{G}$, let $\mathcal{F} = \{(\mathbf{x}, y) \to \ell(f(\mathbf{x}), y)\}$ be a loss function class consisting of the bounded functions with the range $[a, b]$ for a space of labeling functions $\mathcal{G}$. If the following holds*

$$\lim_{m\to\infty} \frac{\ln \mathcal{N}_1(\xi'/8,\mathcal{F},2m)}{\frac{m}{32(b-a)^2}} < \infty,$$

with $\xi' = \xi - D_{\mathcal{F}}(\mathcal{S},\mathcal{T})$, then we have for any $\xi \geq D_{\mathcal{F}}(\mathcal{S},\mathcal{T})$,

$$\lim_{m\to\infty} \mathbf{Pr}\,\{\sup_{f\in\mathcal{F}} |\mathrm{R}^{\ell}_{\hat{\mathcal{S}}} f - \mathrm{R}^{\ell}_{\mathcal{T}} f| > \xi\} = 0.$$

One may note that here the probability of event $\{\sup_{f\in\mathcal{F}} |\mathrm{R}^{\ell}_{\hat{\mathcal{S}}} f - \mathrm{R}^{\ell}_{\mathcal{T}} f| > \xi\}$ is taken with respect to the threshold $\xi \geq D_{\mathcal{F}}(\mathcal{S},\mathcal{T})$, while in standard learning theory this guarantee is usually stated for any $\xi > 0$ given that $\lim_{m\to\infty} \frac{\ln \mathcal{N}_1(\xi,\mathcal{F},m)}{m} < \infty$. This highlights an important difference between the classic generalization bounds for supervised learning and the result given in theorem 5.1.

As we mentioned above, the general setting for generalization bounds with IPMs proposed by Zhang *et al.* suffers from two major drawbacks: (1) the function class in the definition of the IPM is not specified, making it intractable to compute; (2) the proposed bounds are established for joint distributions, rather than marginal distributions, making them not very informative in practice. To this end, we present below two different lines of research that tackle these drawbacks and establish the generalization bounds for domain adaptation by explicitly considering a particular function class with a divergence term taking into account the discrepancy between the marginal distributions of source and target domains. These lines lead to two important particular cases of IPMs that were used to derive generalization bounds in domain adaptation that are the maximum mean discrepancy distance (MMD) and the Wasserstein distance. We take a closer look at both of them in the following sections.

5.3. Wasserstein distance

Despite many important theoretical insights presented before, the above-mentioned divergence measures such as the $\mathcal{H}\Delta\mathcal{H}$-divergence and the discrepancy do not directly take into account the geometry of the data distribution when estimating the discrepancy between two domains. Recently, Courty *et al.* [COU 17] proposed to tackle this drawback by solving the domain adaptation problem using ideas from optimal transportation (OT) theory.

OT theory was first introduced in [MON 81] to study the problem of resource allocation. Assuming that we have a set of factories and a set of mines, the goal of OT is to move the ore from mines to factories in an optimal way, i.e. by minimizing the overall transport cost. More formally, let $\mathbf{X} \subseteq \mathbb{R}^d$ be a measurable space and denote by $\mathcal{P}(\mathbf{X})$ the set of all probability measures over $\mathbf{X}$. Given two probability measures $\mathcal{S}_\mathbf{X}, \mathcal{T}_\mathbf{X} \in \mathcal{P}(\mathbf{X})$, the Monge–Kantorovich problem consists of finding a probabilistic coupling γ defined as a joint probability measure over $\mathbf{X} \times \mathbf{X}$ with marginals $\mathcal{S}_\mathbf{X}$ and $\mathcal{T}_\mathbf{X}$ for all $\mathbf{x}, \mathbf{x}' \in \mathbf{X}$ that minimizes the cost of transport w.r.t. some function $c : \mathbf{X} \times \mathbf{X} \to \mathbb{R}_+$, i.e.

$$\mathop{\arg\min}_{\gamma\in\Pi(\mathcal{S}_\mathbf{X},\mathcal{T}_\mathbf{X})} \int_{\mathbf{X}\times\mathbf{X}} c(\mathbf{x},\mathbf{x}')^p d\gamma(\mathbf{x},\mathbf{x}'),$$

where $\Pi(\mathcal{S}_\mathbf{X}, \mathcal{T}_\mathbf{X})$ is a collection of all joint probability measures on $\mathbf{X} \times \mathbf{X}$ with marginals $\mathcal{S}_\mathbf{X}$ and $\mathcal{T}_\mathbf{X}$. This problem admits a unique solution γ_0 thay allows us to define the Wasserstein distance of order p between $\mathcal{S}_{\mathbf{X}}$ and $\mathcal{T}_{\mathbf{X}}$ for any $p \in [1; +\infty]$ as follows:

$$W_p^p(\mathcal{S}_{\mathbf{X}}, \mathcal{T}_{\mathbf{X}}) = \inf_{\gamma \in \Pi(\mathcal{S}_{\mathbf{X}}, \mathcal{T}_{\mathbf{X}})} \int_{\mathbf{X} \times \mathbf{X}} c(\mathbf{x}, \mathbf{x}')^p d\gamma(\mathbf{x}, \mathbf{x}'),$$

where $c : \mathbf{X} \times \mathbf{X} \to \mathbb{R}_+$ is a cost function for transporting one unit of mass $\mathbf{x}$ to $\mathbf{x}'$.

REMARK 5.1.– *In what follows, we only consider the case $p = 1$ but all the obtained results can be easily extended to the case $p > 1$ using the Hölder inequality, implying for every $p \leq q \Rightarrow W_p \leq W_q$.*

In the discrete case, when one deals with empirical measures

$$\hat{\mathcal{S}}_{\mathbf{X}} = \frac{1}{N_S} \sum_{i=1}^{N_S} \delta_{\mathbf{x}_i^S} \quad \text{and} \quad \hat{\mathcal{T}}_{\mathbf{X}} = \frac{1}{N_T} \sum_{i=1}^{N_T} \delta_{\mathbf{x}_i^T},$$

represented by the uniformly weighted sums of Diracs with mass at locations $S_u = \{\mathbf{x}_i^S\}_{i=1}^{N_S}$ and $T_u = \{\mathbf{x}_i^T\}_{i=1}^{N_T}$, respectively, the Monge–Kantorovich problem is defined in terms of the inner product between the coupling matrix γ and the cost matrix C:

$$W_1(\hat{\mathcal{S}}_{\mathbf{X}}, \hat{\mathcal{T}}_{\mathbf{X}}) = \min_{\gamma \in \Pi(\hat{\mathcal{S}}_{\mathbf{X}}, \hat{\mathcal{T}}_{\mathbf{X}})} \langle C, \gamma \rangle_F,$$

where $\langle \cdot, \cdot \rangle_F$ is the Frobenius dot product, $\Pi(\hat{\mathcal{S}}_{\mathbf{X}}, \hat{\mathcal{T}}_{\mathbf{X}}) = \{\gamma \in \mathbb{R}_+^{N_S \times N_T} | \gamma \mathbf{1} = \hat{\mathcal{S}}_{\mathbf{X}}, \gamma^T \mathbf{1} = \hat{\mathcal{T}}_{\mathbf{X}}\}$ is a set of doubly stochastic matrices and C is a dissimilarity matrix, i.e. $C_{ij} = c(\mathbf{x}_i^S, \mathbf{x}_j^T)$, defining the energy needed to move a probability mass from $\mathbf{x}_i^S$ to $\mathbf{x}_j^T$.

The main underlying idea of Courty *et al.* [COU 17] was to find a coupling matrix that efficiently transports source samples to target ones by solving the following optimization problem:

$$\gamma_o = \operatorname*{argmin}_{\gamma \in \Pi(\hat{\mathcal{S}}_{\mathbf{X}}, \hat{\mathcal{T}}_{\mathbf{X}})} \langle C, \gamma \rangle_F.$$

This idea has a very appealing and intuitive interpretation based on transport from one domain to another. The transportation plan solving OT problem takes into account the geometry of the data by means of an associated cost function, which is based on the Euclidean distance between examples. Furthermore, it is naturally defined as an infimum problem over all feasible solutions. An interesting property of this approach is that the resulting solution given by a joint probability distribution allows one to obtain the new projection of the instances of one domain into another directly without being restricted to a particular hypothesis class. Once the optimal coupling γ_o is found, source samples S_u can be transformed into target-aligned source samples $\hat{S}_u$ using the following equation:

$$\hat{S}_u = \text{diag}\Big((\gamma_o \mathbf{1})^{-1}\Big)\gamma_o T_u.$$

The use of the Wasserstein distance here has an important advantage over other distances used in domain adaptation, as it preserves the topology of the data and admits a rather efficient estimation as mentioned above. Furthermore, as shown in [COU 17], it improves current state-of-the-art results on benchmark computer vision data sets and has very appealing intuition behind it.

In order to justify domain adaptation algorithms based on the minimization of the Wasserstein distance, the generalization bounds for the three domain adaptation settings involving this latter were presented in [RED 17]. According to [VIL 09], the Wasserstein distance is rather strong and can be combined with smoothness bounds to obtain convergences in other distances. As mentioned by the authors, this important advantage of the Wasserstein distance leads to tighter bounds in comparison to other state-of-the-art results and is more computationally attractive as explained below.

5.3.1. *Generalization bounds with the Wasserstein distance*

To introduce the Wasserstein distance into the generalization bounds in domain adaptation scenarios, the authors proposed to consider the following construction. Let $\mathcal{F} = \{f \in \mathcal{H}_k : \|f\|_{\mathcal{H}_k} \leq 1\}$, where $\mathcal{H}_k$ is a reproducing Kernel Hilbert space (RKHS) with its associated kernel k. Let $\ell_{h,f} : \mathbf{x} \to \ell(h(\mathbf{x}), f(\mathbf{x}))$ be a convex loss-function defined-$\forall h, f \in \mathcal{F}$ and assume that ℓ obeys the triangle inequality. As mentioned before, $h(\mathbf{x})$ corresponds to the hypothesis and $f(\mathbf{x})$ to the true labeling functions, respectively. Considering that $(h, f) \in \mathcal{F}^2$, the loss function ℓ is a nonlinear mapping of the RKHS $\mathcal{H}_k$ for the family of ℓ_q losses defined previously[3]. Using results from [SAI 97], one may show that $\ell_{h,f}$ also belongs to the RKHS $\mathcal{H}_{k^q}$, admitting the reproducing kernel k^q and that its norm obeys the following inequality:

$$||\ell_{h,f}||^2_{\mathcal{H}_{k^q}} \leq ||h - f||^{2q}_{\mathcal{H}_k}.$$

This result gives us two important properties of $\ell_{f,h}$ that we use further:

– the function $\ell_{h,f}$ belongs to the RKHS that allows us to use the reproducing property via some feature map $\phi(x)$ associated with kernel k^q;

– the norm $||\ell_{h,f}||_{\mathcal{H}_{k^q}}$ is bounded.

3 If $(h, f) \in \mathcal{F}^2$, then $h - f \in \mathcal{F}$, implying that $\ell(h(\mathbf{x}), f(\mathbf{x})) = |h(\mathbf{x}) - f(\mathbf{x})|^q$ is a nonlinear transform for $h - f \in \mathcal{F}$.

Thus, the error function defined above can be also expressed in terms of the inner product in the corresponding Hilbert space, i.e.[4]

$$\begin{aligned}\mathrm{R}_{\mathcal{S}}^{\ell}(h, f_{\mathcal{S}}) &= \mathop{\mathbf{E}}_{\mathbf{x}\sim\mathcal{S}_{\mathbf{X}}} [\ell(h(\mathbf{x}), f_{\mathcal{S}}(\mathbf{x}))] \\ &= \mathop{\mathbf{E}}_{\mathbf{x}\sim\mathcal{S}_{\mathbf{X}}} [\langle\phi(x), \ell\rangle_{\mathcal{H}_{k^q}}].\end{aligned}$$

The target error can be defined in the same manner

$$\begin{aligned}\mathrm{R}_{\mathcal{T}}^{\ell}(h, f_{\mathcal{T}}) &= \mathop{\mathbf{E}}_{\mathbf{x}\sim\mathcal{T}_{\mathbf{X}}} [\ell(h(\mathbf{x}), f_{\mathcal{T}}(\mathbf{x}))] \\ &= \mathop{\mathbf{E}}_{\mathbf{x}\sim\mathcal{T}_{\mathbf{X}}} [\langle\phi(\mathbf{x}), \ell\rangle_{\mathcal{H}_{k^q}}].\end{aligned}$$

Now, the following lemma that relates the Wasserstein metric with the source and target error functions for an arbitrary pair of hypotheses can be proved.

LEMMA 5.2 ([RED 17]).– *Let $\mathcal{S}_{\mathbf{X}}, \mathcal{T}_{\mathbf{X}} \in \mathcal{P}(\mathbf{X})$ be two probability measures on $\mathbb{R}^d$. Assume that the cost function $c(\mathbf{x}, \mathbf{x}') = \|\phi(\mathbf{x}) - \phi(\mathbf{x}')\|_{\mathcal{H}_{k_\ell}}$, where $\mathcal{H}$ is an RKHS equipped with kernel $k_\ell : \mathbf{X} \times \mathbf{X} \to \mathbb{R}$ induced by $\phi : \mathbf{X} \to \mathcal{H}_{k_\ell}$ and $k_\ell(\mathbf{x}, \mathbf{x}') = \langle\phi(\mathbf{x}), \phi(\mathbf{x}')\rangle_{\mathcal{H}_{k_\ell}}$. Assume further that the loss function $\ell_{h,f} : \mathbf{x} \longrightarrow \ell(h(\mathbf{x}), f(\mathbf{x}))$ is convex, symmetric and bounded and obeys the triangular equality and has the parametric form $|h(\mathbf{x}) - f(\mathbf{x})|^q$ for some $q > 0$. Assume also that kernel k_ℓ in the RKHS $\mathcal{H}_{k_\ell}$ is square-root integrable w.r.t. both $\mathcal{S}_{\mathbf{X}}, \mathcal{T}_{\mathbf{X}}$ for all $\mathcal{S}_{\mathbf{X}}, \mathcal{T}_{\mathbf{X}} \in \mathcal{P}(\mathbf{X})$ where $\mathbf{X}$ is separable and $0 \le k_\ell(\mathbf{x}, \mathbf{x}') \le K, \forall\, \mathbf{x}, \mathbf{x}' \in \mathbf{X}$. If $\|\ell\|_{\mathcal{H}_{k_\ell}} \le 1$, then the following holds:*

$$\forall (h, h') \in \mathcal{H}_{k_\ell}^2, \quad \mathrm{R}_{\mathcal{T}}^{\ell_q}(h, h') \le \mathrm{R}_{\mathcal{S}}^{\ell_q}(h, h') + W_1(\mathcal{S}_{\mathbf{X}}, \mathcal{T}_{\mathbf{X}}).$$

The proof of this lemma can be found in the Appendix 3. This lemma makes use of the Wasserstein distance to relate the source and target errors. The assumption made here is specifying that the cost function $c(\mathbf{x}, \mathbf{x}') = \|\phi(\mathbf{x}) - \phi(\mathbf{x}')\|_{\mathcal{H}}$. While it may seem too restrictive, this assumption is, in fact, not that strong. Using the properties of the inner product, we have:

$$\begin{aligned}\|\phi(\mathbf{x}) - \phi(\mathbf{x}')\|_{\mathcal{H}} &= \sqrt{\langle\phi(\mathbf{x}) - \phi(\mathbf{x}'), \phi(\mathbf{x}) - \phi(\mathbf{x}')\rangle_{\mathcal{H}}} \\ &= \sqrt{k(\mathbf{x}, \mathbf{x}) - 2k(\mathbf{x}, \mathbf{x}') + k(\mathbf{x}, \mathbf{x}')}.\end{aligned}$$

One may further show that for any given positive-definite kernel k, there is a distance c (used as a cost function in our case) that generates it and vice versa (see lemma 12 from [SEJ 13]).

4 For simplicity, we will write ℓ to mean $\ell_{f,h}$.

In order to present the next theorem, we first present an important result showing the convergence of the empirical measure $\hat{\mu}$ to its true associated measure w.r.t. the Wasserstein metric. This concentration guarantee allows us to propose generalization bounds based on the Wasserstein distance for finite samples rather than true measures. It can be specialized for the case of W_1 as follows [BOL 07].

THEOREM 5.3 ([BOL 07], theorem 1.1).– *Let μ be a probability measure in $\mathbb{R}^d$ so that for some $\alpha > 0$, $\int_{\mathbb{R}^d} e^{\alpha\|\mathbf{x}\|^2} d\mu < \infty$, and $\hat{\mu} = \frac{1}{N}\sum_{i=1}^{N} \delta_{\mathbf{x}_i}$ be its associated empirical measure defined on a sample of independent variables $\{\mathbf{x}_i\}_{i=1}^{N}$ drawn from μ. Then for any $d' > d$ and $\varsigma' < \sqrt{2}$, there exists some constant N_0 depending on d' and some square exponential moment of μ such that, for any $\varepsilon > 0$ and $N \geq N_0 \max(\varepsilon^{-(d'+2)}, 1)$,*

$$\mathbf{Pr}\left[W_1(\mu, \hat{\mu}) > \varepsilon\right] \leq \exp\left(-\frac{\varsigma'}{2} N\varepsilon^2\right)$$

where d', ς' can be calculated explicitly.

The convergence guarantee of this theorem can be further strengthened as shown in [FOU 15] but we prefer this version for the ease of reading. We can now use it in combination with the previous lemma to prove the following theorem.

THEOREM 5.4.– *Under the assumptions of lemma 5.2, let S_u and T_u be two samples of size N_S and N_T drawn i.i.d. from $\mathcal{S}_\mathbf{X}$ and $\mathcal{T}_\mathbf{X}$, respectively. Let $\hat{\mathcal{S}}_\mathbf{X} = \frac{1}{N_S}\sum_{i=1}^{N_S} \delta_{x_i^S}$ and $\hat{\mathcal{T}}_\mathbf{X} = \frac{1}{N_T}\sum_{i=1}^{N_T} \delta_{x_i^T}$ be the associated empirical measures. Then for any $d' > d$ and $\varsigma' < \sqrt{2}$, there exists some constant N_0 depending on d' such that for any $\delta > 0$ and $\min(N_S, N_T) \geq N_0 \max(\delta^{-(d'+2)}, 1)$ with probability at least $1 - \delta$ for all h, we have*

$$\mathrm{R}_{\mathcal{T}}^{\ell_q}(h) \leq \mathrm{R}_{\mathcal{S}}^{\ell_q}(h) + W_1(\hat{\mathcal{S}}_\mathbf{X}, \hat{\mathcal{T}}_\mathbf{X}) + \sqrt{2\log\left(\frac{1}{\delta}\right)/\varsigma'}\left(\sqrt{\frac{1}{N_S}} + \sqrt{\frac{1}{N_T}}\right) + \lambda,$$

where λ is the combined error of the ideal hypothesis h^* that minimizes the combined error of $\mathrm{R}_{\mathcal{S}}^{\ell_q}(h) + \mathrm{R}_{\mathcal{T}}^{\ell_q}(h)$.

Given lemma 5.2 and theorem 5.3, the proof of this theorem can be straightforwardly derived in a way similar to that of theorem 3.2 and thus we omit it in this book. A first immediate consequence of this theorem is that it justifies the use of the OT in a domain adaptation context. This does not suggest, however, that minimization of the Wasserstein distance can be done independently of the minimization of the source error, nor does it say that the joint error given by the lambda term becomes small. First, it is clear that the result of W_1 minimization provides a transport of the source to the target such that W_1 becomes small when computing the distance between transported sources and target instances. Under the hypothesis that class labeling is preserved by transport, i.e. $\mathcal{S}(Y|\mathbf{X}) = \mathcal{T}(Y|\text{Transport}(\mathbf{x}))$, the adaptation can be made possible by minimizing W_1 only. However, this is not a reasonable assumption in practice. Indeed, by minimizing the W_1 distance only, it is possible that the obtained transformation transports one positive and one negative source instance to the same target point and then the empirical source error cannot be properly minimized. Additionally, the joint error will be affected since no classifier will be able to separate these source points. We can also think of an extreme case where the positive source examples are transported to negative target instances; in that case the joint error λ will be dramatically affected. A solution is then to regularize the transport to help the minimization of the source error, which can be seen as a kind of joint optimization. This idea was partially implemented as a class-labeled regularization term added to the original optimal transport formulation in [COU 17] and showed good empirical results in practice. The proposed regularized optimization problem reads

$$\min_{\gamma \in \Pi(\hat{\mathcal{S}}_{\mathbf{X}}, \hat{\mathcal{T}}_{\mathbf{X}})} \langle C, \gamma \rangle_F - \frac{1}{\lambda} E(\gamma) + \eta \sum_j \sum_{\mathcal{L}} \|\gamma(I_{\mathcal{L}}, j)\|_q^p.$$

Here, the second term $E(\gamma) = -\sum_{i,j}^{N_S, N_T} \gamma_{i,j} \log(\gamma_{i,j})$ is the regularization term that allows us to solve OT problem efficiently using the Sinkhorn–Knopp matrix scaling algorithm [SIN 67]. The second regularization term $\eta \sum_j \sum_c \|\gamma(I_c, j)\|_q^p$ is used to restrict source examples of different classes to be transported to the same target examples by promoting group sparsity in the matrix γ because of $\| \cdot \|_q^p$ with $q = 1$ and $p = \frac{1}{2}$. In some way, this regularization term influences the capability term by ensuring the existence of a good hypothesis that will be able to be discriminant on both source and target domain data. Another recent paper [PER 16] also suggests that transport regularization is important for the use of OT in domain adaptation tasks. Thus, we conclude that the regularized transport formulations such as that of Courty *et al.* [COU 17] can be seen as algorithmic solutions for controlling the trade-off between the terms of the bound.

5.3.2. *Combined error and semi-supervised case*

To remain consistent with the previous chapters, we also provide the generalization bound for the Wasserstein distance in the semi-supervised setting below.

THEOREM 5.5 ([RED 17]).– *Let S_u, T_u be unlabeled samples of size N_S and N_T each, drawn independently from $\mathcal{S}_{\mathbf{X}}$ and $\mathcal{T}_{\mathbf{X}}$, respectively. Let S be a labeled sample of size m generated by drawing βm points from $\mathcal{T}_{\mathbf{X}}$ ($\beta \in [0,1]$) and $(1 - \beta)m$ points from $\mathcal{S}_{\mathbf{X}}$ and labeling them according to f_S and f_T, respectively. If $\hat{h} \in \mathcal{H}$*

is the empirical minimizer of $\mathrm{R}^{\alpha}_{\hat{S}}(h)$ *on* S *and* $h^*_T = \underset{h \in \mathcal{H}}{\operatorname{argmin}}\ \mathrm{R}^{\ell_q}_{\mathcal{T}}(h)$*, then for any* $\delta \in (0,1)$*, with probability at least* $1-\delta$ *(over the choice of samples), we have*

$$\mathrm{R}^{\ell_q}_{\mathcal{T}}(\hat{h}) \ \leq \ \mathrm{R}^{\ell_q}_{\mathcal{T}}(h^*_T) + c_1 + 2(1-\alpha)(W_1(\hat{\mathcal{S}}_{\mathbf{X}}, \hat{\mathcal{T}}_{\mathbf{X}}) + \lambda + c_2),$$

where

$$c_1 = 2\sqrt{\frac{2K\left(\frac{(1-\alpha)^2}{1-\beta} + \frac{\alpha^2}{\beta}\right)\log(2/\delta)}{m}} + 4\sqrt{K/m}\left(\frac{\alpha}{m\beta\sqrt{\beta}} + \frac{(1-\alpha)}{m(1-\beta)\sqrt{1-\beta}}\right),$$

$$c_2 = \sqrt{2\log\left(\frac{1}{\delta}\right)/\varsigma'}\left(\sqrt{\frac{1}{N_S}} + \sqrt{\frac{1}{N_T}}\right).$$

The proof of this theorem is given in Appendix 3. In line with the results obtained previously, this theorem shows that the best hypothesis that takes into account both source and target labeled data (i.e. $0 \leq \alpha < 1$) performs at least as well as the best hypothesis learned on target data instances alone ($\alpha = 1$). This result agrees well with the intuition that semi-supervised domain adaptation approaches should be at least as good as unsupervised ones.

5.4. Maximum mean discrepancy distance

Maximum mean discrepancy plays a very important role in domain adaptation as over the years it has been widely used to derive new efficient domain adaptation algorithms. These algorithms usually proceeded by finding a weight vector that minimized the above-mentioned discrepancy between source and target samples. Initially proposed and studied in the context of kernel two-sample tests in statistics, maximum mean discrepancy has further gained popularity due to the existence of several efficient estimators and algorithms that can be used to calculate and minimize it from finite samples.

In this section, we briefly introduce kernel embeddings of distribution functions. We start with a definition of a mean map and its empirical estimate.

DEFINITION 5.2.– *Let* $k : \mathcal{X} \times \mathcal{X} \to \mathbb{R}$ *be a kernel in the RKHS* $\mathcal{H}_k$ *and let* $\phi(\mathbf{x}) = k(\mathbf{x}, \cdot)$ *be its associated feature map. Given a probability distribution* $\mathcal{D}_{\mathbf{X}}$*, the mapping*

$$\mu[\mathcal{D}_{\mathbf{X}}] = \underset{\mathbf{x} \sim \mathcal{D}_{\mathbf{X}}}{\mathbf{E}}[\phi(\mathbf{x})]$$

is called a mean map. Its empirical value is given by the following estimate:

$$\mu[X] = \frac{1}{m}\sum_{i=1}^{m}\phi(\mathbf{x}_i),$$

where $X = \{\mathbf{x}_1, \ldots, \mathbf{x}_m\}$ *is drawn i.i.d. from* $\mathcal{D}_{\mathbf{X}}$.

If $\mathbf{E}_{\mathbf{x}\sim\mathcal{D}_{\mathbf{X}}}\left[k(\mathbf{x},\mathbf{x})\right] < \infty$, then $\mu[\mathcal{D}_{\mathbf{X}}]$ is an element of RKHS $\mathcal{H}_k$. According to the Moore–Aronszajn theorem, the reproducing property of $\mathcal{H}_k$ allows us to rewrite every function $f \in \mathcal{H}_k$ in the following form: $\langle \mu[\mathcal{D}_{\mathbf{X}}], f\rangle = \mathbf{E}_{\mathbf{x}\sim\mathcal{D}_{\mathbf{X}}}\left[f(\mathbf{x})\right]$. Then, the MMD distance is defined as follows.

DEFINITION 5.3 (Maximum mean discrepancy).– *Let* $\mathcal{F} = \{f \in \mathcal{H}_k : \|f\|_{\mathcal{H}_k} \leq 1\}$ *where* $\mathcal{H}_k$ *is an RKHS with its associated kernel k. Let* $\mathcal{S}_{\mathbf{X}}$ *and* $\mathcal{T}_{\mathbf{X}}$ *be two probability Borel measures. Then we define* $d_{MMD}(\mathcal{S}_{\mathbf{X}}, \mathcal{T}_{\mathbf{X}})$ *as*

$$d_{MMD}(\mathcal{S}_{\mathbf{X}}, \mathcal{T}_{\mathbf{X}}) = \sup_{f\in\mathcal{F}}\left[\mathop{\mathbf{E}}_{x\sim\mathcal{S}_{\mathbf{X}}}[f(x)] - \mathop{\mathbf{E}}_{y\sim\mathcal{T}_{\mathbf{X}}}[f(y)]\right].$$

This expression can be further simplified by using the definition of a kernel embedding of a probability distribution given above. We write:

$$d_{\mathrm{MMD}}(\mathcal{S}_{\mathbf{X}}, \mathcal{T}_{\mathbf{X}}) = \|\mu[\mathcal{S}_{\mathbf{X}}] - \mu[\mathcal{T}_{\mathbf{X}}]\|_{\mathcal{H}_k}.$$

From practical point of view, we observe that numerous domain adaptation and transfer learning approaches are based on MMD minimization [PAN 09, GEN 11, HUA 06, PAN 08, CHE 09]. Furthermore, conditional kernel embeddings proved to be efficient for target and conditional shift correction [ZHA 13] and vector-valued regression [GRÜ 12]. Thus, the use of this metric in domain adaptation theory is justified and appears to be natural.

5.4.1. *Generalization bound with kernel embeddings*

Based on the results with the Wasserstein distance, we now introduce generalization bounds for the source and target error where the divergence between tasks' distributions is measured by the MMD distance. As mentioned before, we start with a lemma that relates the source and target error in terms of the introduced discrepancy measure for an arbitrary pair of hypotheses. Then, we show how target error can be bounded by the empirical estimate of the MMD plus the complexity term.

LEMMA 5.3 ([RED 15]).– *Let $\mathcal{F} = \{f \in \mathcal{H}_k : \|f\|_{\mathcal{H}_k} \leq 1\}$, where $\mathcal{H}_k$ is an RKHS with its associated kernel k. Let $\ell_{h,f} : \mathbf{x} \to \ell(h(\mathbf{x}), f(\mathbf{x}))$ be a convex loss-function with parametric form $|h(\mathbf{x}) - f(\mathbf{x})|^q$ for some $q > 0$ and defined $\forall h, f \in \mathcal{F}$ such that ℓ obeys the triangle inequality. Then, if $\|l\|_{\mathcal{H}_{k^q}} \leq 1$, we have:*

$$\forall (h, h') \in \mathcal{F}, \quad \mathrm{R}_{\mathcal{T}}^{\ell_q}(h, h') \leq \mathrm{R}_{\mathcal{S}}^{\ell_q}(h, h') + d_{MMD}(\mathcal{S}_{\mathbf{X}}, \mathcal{T}_{\mathbf{X}}).$$

This lemma is proved in a similar way to lemma 5.2 from [RED 17] presented earlier in this chapter. Using this and the result that relates the true and empirical MMD distance [SON 08], we can prove the following theorem.

THEOREM 5.6.– *With the assumptions from lemma 5.3, let S_u and T_u be two samples of size m drawn i.i.d. from $\mathcal{S}$ and $\mathcal{T}$, respectively. Then, with probability at least $1 - \delta(\delta \in (0,1))$ for all $h \in \mathcal{F}$ the following holds:*

$$\begin{aligned}\mathrm{R}_{\mathcal{T}}^{\ell_q}(h) &\leq \mathrm{R}_{\mathcal{S}}^{\ell_q}(h) + d_{MMD}(\hat{\mathcal{S}}_{\mathbf{X}}, \hat{\mathcal{T}}_{\mathbf{X}}) \\ &\quad + \frac{2}{m}\left(\mathop{\mathbf{E}}_{x \sim \mathcal{S}_{\mathbf{X}}}\left[\sqrt{\mathrm{tr}(K_{\mathcal{S}})}\right] + \mathop{\mathbf{E}}_{x \sim \mathcal{T}_{\mathbf{X}}}\left[\sqrt{\mathrm{tr}(K_{\mathcal{T}})}\right]\right) + 2\sqrt{\frac{\log(\frac{2}{\delta})}{2m}} + \lambda,\end{aligned}$$

where $d_{MMD}(\hat{\mathcal{S}}_{\mathbf{X}}, \hat{\mathcal{T}}_{\mathbf{X}})$ is an empirical counterpart of $d_{MMD}(\mathcal{S}_{\mathbf{X}}, \mathcal{T}_{\mathbf{X}})$, $K_{\mathcal{S}}$ and $K_{\mathcal{T}}$ are the kernel functions calculated on samples from $\mathcal{S}_{\mathbf{X}}$ and $\mathcal{T}_{\mathbf{X}}$, respectively, and λ is the combined error of the ideal hypothesis h^* that minimizes the combined error of $\mathrm{R}_{\mathcal{S}}^{\ell_q}(h) + \mathrm{R}_{\mathcal{T}}^{\ell_q}(h)$.

We can see that this theorem is similar in shape to theorem 5.4 and theorem 3.2. The main difference, however, is that the complexity term does not depend on the Vapnik–Chervonenkis dimension. In our case, the loss function between two errors is bounded by the empirical MMD between distributions and two terms that correspond to the empirical Rademacher complexities of $\mathcal{H}$ w.r.t. the source and target samples. In both theorems, λ plays the role of the combined error of the ideal hypothesis. Its presence in the bound comes from the use of the triangle inequality for classification error.

This result is particularly useful as squared MMD distance $d^2_{\text{MMD}}(\hat{\mathcal{S}}_{\mathbf{X}}, \hat{\mathcal{T}}_{\mathbf{X}})$ can be calculated as

$$d^2_{\text{MMD}}(\hat{\mathcal{S}}_{\mathbf{X}}, \hat{\mathcal{T}}_{\mathbf{X}}) = \frac{1}{m(m-1)} \sum_{i \neq j} h((\mathbf{x}_i, \mathbf{x}_j), (\mathbf{y}_i, \mathbf{y}_j)),$$

where for $\mathbf{x} \in S_u$ and $\mathbf{y} \in T_u$ we have

$$h((\mathbf{x}_i, \mathbf{x}_j), (\mathbf{y}_i, \mathbf{y}_j)) = \tfrac{1}{m^2} \sum_{i,j=1}^{m} k(\mathbf{x}_i, \mathbf{x}_j) - \tfrac{2}{m^2} \sum_{i,j=1}^{m} k(\mathbf{x}_i, \mathbf{y}_j) + \tfrac{1}{m^2} \sum_{i,j=1}^{m} k(\mathbf{y}_i, \mathbf{y}_j).$$

The following lemma gives a computation guarantee for the unbiased estimator of $d^2_{\text{MMD}}(\hat{\mathcal{S}}_{\mathbf{X}}, \hat{\mathcal{T}}_{\mathbf{X}})$.

LEMMA 5.4 ([GRE 12]).– *For $m_2 = m/2$, the estimator*

$$d^2_{MMD}(\hat{\mathcal{S}}_{\mathbf{X}}, \hat{\mathcal{T}}_{\mathbf{X}}) = \frac{1}{m_2} \sum_{i=1}^{m_2} h((\mathbf{x}_{2i-1}, \mathbf{y}_{2i-1})(\mathbf{x}_{2i}, \mathbf{y}_{2i}))$$

can be computed in linear time and is an unbiased estimator of $d^2_{MMD}(\mathcal{S}_{\mathbf{X}}, \mathcal{T}_{\mathbf{X}})$.

We also note that the obtained bound can be further simplified if one uses, for instance, Gaussian, exponential or Laplacian kernels to calculate kernel functions $K_{\mathcal{S}}$ and $K_{\mathcal{T}}$ as they have 1s on the diagonal, thus facilitating the calculation of the trace. Finally, it can be seen that the bound from theorem 5.6 has the same terms as theorem 3.2 while the MMD distance is estimated as in corollary 3.1.

5.4.2. *Combined error and semi-supervised case*

Similar to the case considered in [BEN 10a], one can also derive similar bounds for the MMD distance in the case of combined error. To this end, we present the following analog of theorem 3.3.

THEOREM 5.7.– *With the assumptions from lemma 5.3, let S_u, T_u be unlabeled samples of size m', each drawn independently from $\mathcal{S}_{\mathbf{X}}$ and $\mathcal{T}_{\mathbf{X}}$, respectively. Let S be a labeled sample of size m generated by drawing $\beta\, m$ points from $\mathcal{T}$ ($\beta \in [0,1]$) and $(1-\beta)\, m$ points from $\mathcal{S}$ and labeling them according to $f_{\mathcal{S}}$ and $f_{\mathcal{T}}$, respectively. If $\hat{h} \in \mathcal{H}$ is the empirical minimizer of $\mathrm{R}^{\alpha}(h)$ on S and $h^*_{\mathcal{T}} = \operatorname*{argmin}_{h\in\mathcal{H}} \mathrm{R}^{\ell_q}_{\mathcal{T}}(h)$, then for any $\delta \in (0,1)$, with probability at least $1-\delta$ (over the choice of samples),*

$$\mathrm{R}^{\ell_q}_{\mathcal{T}}(\hat{h}) \leq \mathrm{R}^{\ell_q}_{\mathcal{T}}(h^*_{\mathcal{T}}) + c_1 + c_2,$$

$$c_1 = 2\sqrt{\frac{2K\left(\frac{(1-\alpha)^2}{1-\beta} + \frac{\alpha^2}{\beta}\right)\log\frac{2}{\delta}}{m}} + 2\left(\sqrt{\frac{\alpha}{\beta}} + \sqrt{\frac{1-\alpha}{1-\beta}}\right)\sqrt{\frac{K}{m}},$$

$$c_2 = \hat{d}_{MMD}(S_u, T_u) + \frac{2}{m'} \mathop{\mathbf{E}}_{\mathbf{x}\sim\mathcal{S}_{\mathbf{X}}} \sqrt{\mathrm{tr}(K_{\mathcal{S}})} + \frac{2}{m'} \mathop{\mathbf{E}}_{\mathbf{x}\sim\mathcal{T}_{\mathbf{X}}} \sqrt{\mathrm{tr}(K_{\mathcal{T}})} + 2\sqrt{\frac{\log\frac{2}{\delta}}{2m'}} + \lambda.$$

Several observations can be made from this theorem. First of all, the main quantities that define the potential success of domain adaptation according to [BEN 10a] (i.e. the distance between the distributions and the combined error of the joint ideal hypothesis) are preserved in the bound. This is an important point that indicates that the two results are not contradictory or supplementary. Second, rewriting the approximation of the bound as a function of α and omitting additive constants can lead to a similar result to in theorem 3.3. This observation may point out the existence of a strong connection between them.

The generalization guarantees obtained for domain adaptation based on MMD distance allow us to take another step forward in domain adaptation theory by extending the results presented in previous sections in two different ways. Similar to discrepancy-based results, the bounds with MMD distance allow us to consider any arbitrary loss function so that not only binary but any multi-class classification application of domain adaptation can be studied. On the other hand, MMD distance, similar to the Sinkhorn distance, has some very nice estimation guarantees that are unavailable for both $\mathcal{H}\Delta\mathcal{H}$ and $disc$ divergences. This feature can be very important in accessing both the *a priori* hardness of adaptation and its *a posteriori* success in order to understand if a given adaptation algorithm manages to properly reduce the discrepancy between domains.

5.5. Relationship between the MMD and the Wasserstein distances

At this point, we have presented two results that introduced the MMD and the Wasserstein distances to the DA generalization bounds for both semi-supervised and unsupervised cases. As both results are built on the same construction, one may want to explore the link between the MMD distance and the Wasserstein distance. In order to do this, we first observe that in some particular cases the MMD distance can be bounded by the Wasserstein metric. Indeed, if we assume that the ground metric in the Wasserstein distance is equal to $c(x, y) = \|\phi(x) - \phi(y)\|_{\mathcal{H}}$, the following results can be obtained:

$$\begin{aligned}
\left\| \int_{\mathbf{X}} f d(\mathcal{S}_{\mathbf{X}} - \mathcal{T}_{\mathbf{X}}) \right\|_{\mathcal{H}} &= \left\| \int_{\mathbf{X}\times\mathbf{X}} (f(x) - f(y)) d\gamma(x, y) \right\|_{\mathcal{H}} \\
&\leq \int_{\mathbf{X}\times\mathbf{X}} \|f(x) - f(y)\|_{\mathcal{H}} d\gamma(x, y) \\
&= \int_{\mathbf{X}\times\mathbf{X}} \| \langle f(x), \phi(x) \rangle - \langle f(y), \phi(y) \rangle \|_{\mathcal{H}} d\gamma(x, y) \\
&\leq \|f\|_{\mathcal{H}} \int_{\mathbf{X}\times\mathbf{X}} \|\phi(x) - \phi(y)\|_{\mathcal{H}} d\gamma(x, y).
\end{aligned}$$

Now taking the supremum over f w.r.t. $\mathcal{F} = \{f : \|f\|_{\mathcal{H}} \leq 1\}$ and infimum over $\gamma \in \Pi(\mathcal{S}_{\mathbf{X}}, \mathcal{T}_{\mathbf{X}})$ gives

$$d_{\text{MMD}}(\mathcal{S}_{\mathbf{X}}, \mathcal{T}_{\mathbf{X}}) \leq W_1(\mathcal{S}_{\mathbf{X}}, \mathcal{T}_{\mathbf{X}}). \quad [5.2]$$

This result holds under the hypothesis that $c(x, y) = \|\phi(x) - \phi(y)\|_{\mathcal{H}}$. On the other hand, in [GAO 14], the authors showed that $W_1(\mathcal{S}_{\mathbf{X}}, \mathcal{T}_{\mathbf{X}})$ with this particular ground metric can be further bounded as follows:

$$W_1(\mathcal{S}_{\mathbf{X}}, \mathcal{T}_{\mathbf{X}}) \leq \sqrt{d^2_{\text{MMD}}(\mathcal{S}_{\mathbf{X}}, \mathcal{T}_{\mathbf{X}}) + C},$$

where $C = \|\mu[\mathcal{S}_{\mathbf{X}}]\|_{\mathcal{H}} + \|\mu[\mathcal{T}_{\mathbf{X}}]\|_{\mathcal{H}}$. This result is quite strong for multiple reasons. First, it allows us to introduce the squared MMD distance to the domain adaptation bounds using [RED 17, lemma 1], leading to the following result for two arbitrary hypotheses $(h, h') \in \mathcal{H}^2$:

$$\mathrm{R}_{\mathcal{T}}(h, h') \leq \mathrm{R}_{\mathcal{S}}(h, h') + \sqrt{d^2_{\text{MMD}}(\mathcal{S}_{\mathbf{X}}, \mathcal{T}_{\mathbf{X}}) + C}.$$

On the other hand, the unified inequality

$$d_{\text{MMD}}(\mathcal{S}_{\mathbf{X}}, \mathcal{T}_{\mathbf{X}}) \leq W_1(\mathcal{S}_{\mathbf{X}}, \mathcal{T}_{\mathbf{X}}) \leq \sqrt{d^2_{\text{MMD}}(\mathcal{S}_{\mathbf{X}}, \mathcal{T}_{\mathbf{X}}) + \|\mu[\mathcal{S}_{\mathbf{X}}]\|_{\mathcal{H}} + \|\mu[\mathcal{T}_{\mathbf{X}}]\|_{\mathcal{H}}} \quad [5.3]$$

suggests that the MMD distance establishes an interval bound for the Wasserstein distance. This point is very interesting because originally the calculation of the Wasserstein distance (also known as the Earth Mover's distance) requires one to solve a linear programming problem, which can be quite time consuming due to the computational complexity of $\mathcal{O}(n^3 \log(n))$ where n is the number of instances.

This result, however, is true only under the assumption that $c(\mathbf{x}, \mathbf{x}') = \|\phi(\mathbf{x}) - \phi(\mathbf{x}')\|_{\mathcal{H}}$. While in most applications the Euclidean distance $c(\mathbf{x}, \mathbf{x}') = \|\mathbf{x} - \mathbf{x}'\|$ is used as a ground metric, this assumption can represent an important constraint. Luckily, it can be circumvented due to the duality between the RKHS-based and distance-based metric representations studied in [SEJ 13]. Let us first rewrite the ground metric as

$$\begin{aligned}\|\phi(\mathbf{x}) - \phi(\mathbf{x}')\|_{\mathcal{H}} &= \sqrt{\langle \phi(\mathbf{x}) - \phi(\mathbf{x}'), \phi(\mathbf{x}) - \phi(\mathbf{x}')\rangle_{\mathcal{H}}} \\ &= \sqrt{k(\mathbf{x}, \mathbf{x}) - 2k(\mathbf{x}, \mathbf{x}') + k(\mathbf{x}', \mathbf{x}')}.\end{aligned}$$

Now, in order to obtain the standard Euclidean distance in the expression of the ground metric, we can pick a kernel given by the covariance function of the fractional Brownian motion, i.e. $k(\mathbf{x}, \mathbf{x}') = \frac{1}{2}(\|\mathbf{x}\|^2 + \|\mathbf{x}'\|^2 - 2\|\mathbf{x} - \mathbf{x}'\|^2)$. Plugging this expression into the definition of $c(\mathbf{x}, \mathbf{x}')$ gives the desired Euclidean distance and thus allows us to calculate the Wasserstein distance with the standard ground metric. We now move on to the unifying inequality that compares multiple metrics and that has been used in the theoretical analysis of the DA problem.

To empirically verify this inequality, we generated two 2D data sets sampled from the Gaussian distributions with means $\left(\begin{smallmatrix}0\end{smallmatrix}\right)$ and $\left(\begin{smallmatrix}4\end{smallmatrix}\right)$ and covariance matrices $\left(\begin{smallmatrix}1 & 0\\ 0 & 1\end{smallmatrix}\right)$, and $\left(\begin{smallmatrix}1 & -.8\\ -.8 & 1\end{smallmatrix}\right)$, and varied their size from 50 to 20,000 data points. We plot the values of $\hat{d}_{\mathrm{MMD}}(\hat{\mathcal{S}}_{\mathbf{X}}, \hat{\mathcal{T}}_{\mathbf{X}})$, $W_1(\hat{\mathcal{S}}_{\mathbf{X}}, \hat{\mathcal{T}}_{\mathbf{X}})$ and $\sqrt{d^2_{\mathrm{MMD}}(\hat{\mathcal{S}}_{\mathbf{X}}, \hat{\mathcal{T}}_{\mathbf{X}}) + \|\mu[S_u]\|_{\mathcal{H}} + \|\mu[T_u]\|_{\mathcal{H}}}$.

For fair comparison, we used the Euclidean distance to compute the Wasserstein distance and the covariance function of the fractional Brownian motion as a kernel to compute the MMD. Furthermore, we approximated the empirical embeddings of both samples based on the feature map corresponding to this kernel. The results of this comparison are presented in Figure 5.1. From it, we can see that the MMD distance provides a close lower bound on the Wasserstein distance showing the left-hand side inequality in equation [5.2]. On the other hand, the upper bound on the Wasserstein distance appears to be more vacuous, probably due to the use of the inequality $-2\int_{\mathbf{X}\times\mathbf{X}} k(\mathbf{x}, \mathbf{x}')d\mathcal{S}_{\mathbf{X}}\mathcal{T}_{\mathbf{X}} \leq \int_{\mathbf{X}\times\mathbf{X}} k(\mathbf{x}, \mathbf{x}')d(\mathcal{S}_{\mathbf{X}} - \mathcal{T}_{\mathbf{X}})^2$ that was applied in [GAO 14] in order to bound the Wasserstein distance by the squared MMD distance.

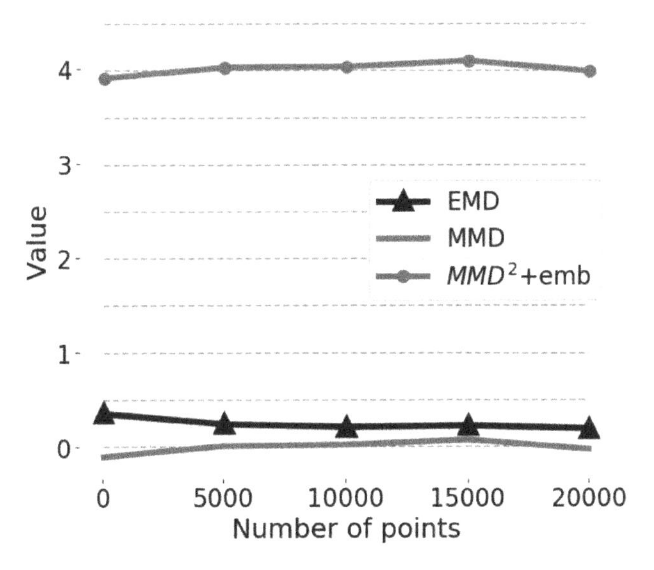

Figure 5.1. *Bound on the Wasserstein distance expressed using the MMD distance. Here, EMD stands for the W_1 distance and MMD^2 + emb is the expression given in equation [5.3]*

5.6. Summary

In this chapter, we presented several theoretical results that use IPMs as a measure of divergence between the marginal source and target domains' distributions in the DA generalization bounds. We argued that this particular choice of distance provides a number of advantages compared to the $\mathcal{H}\Delta\mathcal{H}$ distance and the discrepancy distances considered before. First, both the Wasserstein distance and the MMD distance can be calculated from available finite samples in a computationally attractive way due to the existence of linear time estimators for their entropy-regularized and quadratic versions, respectively. Second, the Wasserstein distance allows us to take geometrical information into account when calculating the divergence between the two domains' distributions, while the MMD distance is calculated based on the distance between the embeddings of two distributions in some (possibly) richer space. This feature is quite interesting as it provides more flexibility when it comes to incorporating the prior knowledge into the DA problem on the one hand, and allows us to have a potentially richer characterization of the divergence between the domains on the other hand. This may explain the abundance of DA algorithms based on the MMD distance and some recent DA techniques developed based on OT theory. Finally, we note that, in general, the presented bounds are similar in shape to those described in Chapter 3 and preserve their main terms, thus remaining consistent with them. This shows that despite a large variety of ways that may be used to formally characterize the generalization phenomenon in domain adaptation, the intuition behind this process and the main factors defining its potential success remain the same.

6

PAC–Bayesian Theory for Domain Adaptation

As recalled in section 1.2.3, the PAC–Bayesian theory [MCA 99] offers tools to derive generalization bounds for models that take the form of a weighted majority vote over a set of functions that can be the hypothesis space. In this section, we recall the work done by Germain *et al.* [GER 16, GER 13] on how the PAC–Bayesian theory can help to theoretically understand domain adaptation through the weighted majority vote learning point of view.

6.1. The PAC–Bayesian framework

In the PAC–Bayesian setting, we recall that we consider a π distribution over the hypothesis set $\mathcal{H}$, and the objective is to learn a ρ distribution over $\mathcal{H}$ by taking into account the information captured by the learning sample S. In the domain adaptation setting, we then aim to learn the ρ-weighted majority vote

$$\forall \mathbf{x} \in \mathbf{X}, \quad B_\rho(\mathbf{x}) = \operatorname{sign}\left[\mathop{\mathbf{E}}_{h\sim\rho} h(\mathbf{x})\right],$$

with the best performance on the target domain $\mathcal{T}$. Note that, here, we consider the $0-1$ loss function. As in a non-adaptation setting, PAC–Bayesian domain adaptation generalization bounds do not directly upper bound $\mathrm{R}_{\mathcal{T}}^{\ell_{01}}(B_\rho)$, but upper bound the expectation according to ρ of the individual risks of the functions from $\mathcal{H}$: $\mathbf{E}_{h\sim\rho}\,\mathrm{R}^{\ell_{01}}(h)$, which is closely related to B_ρ (see equation [1.4]). Let us introduce a tight relation between $\mathrm{R}_{\mathcal{D}}(B_\rho)$ and $\mathbf{E}_{h\sim\rho}\mathrm{R}^{\ell_{01}}(h)$ known as the C-bound (first proved by Lacasse *et al.* [LAC 06]) and defined for all distribution $\mathcal{D}$ on $\mathbf{X}\times Y$ as

$$\mathrm{R}^{\ell_{01}}_{\mathcal{D}}(B_\rho) \;\leq\; 1-\frac{\left(1-2\mathop{\mathbf{E}}_{h\sim\rho}\mathrm{R}^{\ell_{01}}_{\mathcal{D}}(h)\right)^2}{1-2d_{\mathcal{D}_{\mathbf{X}}}(\rho)} \qquad [6.1]$$

where

$$d_{\mathcal{D}_{\mathbf{X}}}(\rho) \;=\; \mathop{\mathbf{E}}_{(h,h')\sim\rho^2}\;\mathop{\mathbf{E}}_{\mathbf{x}\sim\mathcal{D}_{\mathbf{X}}}\ell_{01}\big(h(\mathbf{x}),h'(\mathbf{x})\big)$$

is the expected disagreement between pairs of voters on the marginal distribution $\mathcal{D}_{\mathbf{X}}$. It is important to highlight that the expected disagreement $d_{\mathcal{D}_{\mathbf{X}}}(\rho)$ is closely related to the notion of expected joint error $e_{\mathcal{D}}(\rho)$ between pairs of voters:

$$e_{\mathcal{D}}(\rho) \;=\; \mathop{\mathbf{E}}_{(h,h')\sim\rho^2}\;\mathop{\mathbf{E}}_{(\mathbf{x},y)\sim\mathcal{D}}\ell_{01}\big(h(\mathbf{x}),y)\big)\times\ell_{0-1}\big(h'(\mathbf{x}),y)\big).$$

Indeed, for all distribution $\mathcal{D}$ on $\mathbf{X}\times Y$ we have

$$\mathop{\mathbf{E}}_{h\sim\rho}\mathrm{R}^{\ell_{01}}_{\mathcal{D}}(h) \;=\; \frac{1}{2}d_{\mathcal{D}_{\mathbf{X}}}(\rho)+e_{\mathcal{D}}(\rho). \qquad [6.2]$$

In the following, we recall the two PAC–Bayesian generalization bounds for domain adaptation proposed by Germain *et al.* [GER 13, GER 16] through the point of view of Catoni [CAT 07] (recalled in theorem 1.5 in Chapter 1).

6.2. Two PAC–Bayesian analyses of domain adaptation

There exist two types of bounds for domain adaptation that leads to PAC–Bayesian generalization bounds. The first one [GER 13] is based on a divergence measure that follows the same spirit than the one induced by the first analysis of domain adaptation (see sections 3.1 and 3.5.2). The second one takes advantages of inherent properties of the PAC–Bayesian framework.

6.2.1. *In the spirit of Ben-David* et al. *and Mansour* et al.

Germain *et al.* [GER 13] proposed to define a divergence measure that follows the underlying idea carried by the C-bound of equation [6.1]. Concretly, if $\mathbf{E}_{h\sim\rho}\mathrm{R}^{\ell_{01}}_{\mathcal{S}}(h)$ and $\mathbf{E}_{h\sim\rho}\mathrm{R}^{\ell_{01}}_{\mathcal{T}}(h)$ are similar, then $\mathrm{R}^{\ell_{01}}_{\mathcal{S}}(B_\rho)$ and $\mathrm{R}^{\ell_{01}}_{\mathcal{T}}(B_\rho)$ are similar when $d_{\mathcal{S}_{\mathbf{X}}}(\rho)$ and $d_{\mathcal{T}_{\mathbf{X}}}(\rho)$ are also similar. Thus, the domains $\mathcal{S}$ and $\mathcal{T}$ are close according to ρ if the expected disagreement over the two domains tends to be close. Germain *et al.* [GER 13] introduced the following domain disagreement pseudometric.

DEFINITION 6.1 (Domain disagreement [GER 13]).– *Let $\mathcal{H}$ be a hypothesis class. For any marginal distributions $\mathcal{S}_{\mathbf{X}}$ and $\mathcal{T}_{\mathbf{X}}$ over $\mathbf{X}$, any distribution ρ on $\mathcal{H}$, the domain disagreement* $\operatorname{dis}_\rho(\mathcal{S}_{\mathbf{X}}, \mathcal{T}_{\mathbf{X}})$ *between $\mathcal{S}_{\mathbf{X}}$ and $\mathcal{T}_{\mathbf{X}}$ is defined by*

$$\operatorname{dis}_\rho(\mathcal{S}_{\mathbf{X}}, \mathcal{T}_{\mathbf{X}}) = \Big| d_{\mathcal{T}_{\mathbf{X}}}(\rho) - d_{\mathcal{S}_{\mathbf{X}}}(\rho) \Big| .$$

It is worth noting that the value of $\operatorname{dis}_\rho(\mathcal{S}_{\mathbf{X}}, \mathcal{T}_{\mathbf{X}})$ is always lower than the $\mathcal{H}\Delta\mathcal{H}$-distance between $\mathcal{S}_{\mathbf{X}}$ and $\mathcal{T}_{\mathbf{X}}$. Indeed, for every $\mathcal{H}$ and ρ over $\mathcal{H}$, we have

$$\begin{aligned}
\tfrac{1}{2}\, d_{\mathcal{H}\Delta\mathcal{H}}(\mathcal{S}_{\mathbf{X}}, \mathcal{T}_{\mathbf{X}}) &= \sup_{(h,h')\in\mathcal{H}^2} \left| \mathop{\mathbf{E}}_{\mathbf{x}\sim\mathcal{S}_{\mathbf{X}}} \ell_{01}\big(h(\mathbf{x}), h'(\mathbf{x})\big) - \mathop{\mathbf{E}}_{\mathbf{x}\sim\mathcal{T}} \ell_{01}\big(h(\mathbf{x}), h'(\mathbf{x})\big) \right| \\
&\geq \mathop{\mathbf{E}}_{(h,h')\sim\rho^2} \left| \mathop{\mathbf{E}}_{\mathbf{x}\sim\mathcal{S}_{\mathbf{X}}} \ell_{01}\big(h(\mathbf{x}), h'(\mathbf{x})\big) - \mathop{\mathbf{E}}_{\mathbf{x}\sim\mathcal{T}} \ell_{01}\big(h(\mathbf{x}), h'(\mathbf{x})\big) \right| \\
&\geq \Big| d_{\mathcal{T}_{\mathbf{X}}}(\rho) - d_{\mathcal{S}_{\mathbf{X}}}(\rho) \Big| \\
&= \operatorname{dis}_\rho(\mathcal{S}_{\mathbf{X}}, \mathcal{T}_{\mathbf{X}}).
\end{aligned}$$

From this domain's divergence, the authors proved the following domain adaptation bound.

THEOREM 6.1 ([GER 13]).– *Let $\mathcal{H}$ be a hypothesis class. We have*

$$\forall \rho \text{ on } \mathcal{H},\ \mathop{\mathbf{E}}_{h\sim\rho} \mathrm{R}_{\mathcal{T}}^{\ell_{01}}(h) \leq \mathop{\mathbf{E}}_{h\sim\rho} \mathrm{R}_{\mathcal{S}}^{\ell_{01}}(h) + \frac{1}{2} \operatorname{dis}_\rho(\mathcal{S}_{\mathbf{X}}, \mathcal{T}_{\mathbf{X}}) + \lambda_\rho,$$

where λ_ρ is the deviation between the expected joint errors between pairs for voters on the target and source domains, which is defined as

$$\lambda_\rho = \Big| e_{\mathcal{T}}(\rho) - e_{\mathcal{S}}(\rho) \Big| . \qquad [6.3]$$

The proof of this result can be found in Appendix 4. With the above theorem, one can prove different kinds of PAC–Bayesian generalization bounds (see section 1.2.3). We present below a generalization bound that has the advantage of leading to an algorithm [GER 13].

THEOREM 6.2.– *For any domains $\mathcal{S}$ and $\mathcal{T}$ over $\mathbf{X} \times Y$, any set of voters $\mathcal{H}$, any prior distribution π over $\mathcal{H}$, any $\delta \in (0, 1]$, any real numbers $\omega > 0$ and $a > 0$, with a probability at least $1 - \delta$ over the random choice of $S \times T_u \sim (\mathcal{S} \times \mathcal{T}_{\mathbf{X}})^m$, for every posterior distribution ρ on $\mathcal{H}$, we have*

$$\begin{aligned}
\mathop{\mathbf{E}}_{h\sim\rho} \mathrm{R}_{\mathcal{T}}^{\ell_{01}}(h) \leq{}& \omega' \mathop{\mathbf{E}}_{h\sim\rho} \mathrm{R}_{S}^{\ell_{01}}(h) + a' \tfrac{1}{2} \operatorname{dis}_\rho(S, T_u) \\
&+ \left(\frac{\omega'}{\omega} + \frac{a'}{a} \right) \frac{\mathrm{KL}(\rho|\pi) + \ln\frac{3}{\delta}}{m} + \lambda_\rho + \tfrac{1}{2}(a' - 1),
\end{aligned}$$

where $\mathrm{dis}_\rho(S, T_u)$ *are the empirical estimate of the domain disagreement;* λ_ρ *is defined by equation* [6.3]*;* $\omega' = \frac{\omega}{1-e^{-\omega}}$ *and* $a' = \frac{2a}{1-e^{-2a}}$.

Similar to the bounds of theorems 3.1 and 3.4, this bound can be seen as a trade-off between different quantities. Concretely, the terms $\mathbf{E}_{h\sim\rho} \mathrm{R}_S^{\ell_{01}}(h)$ and $\mathrm{dis}_\rho(S, T)$ are akin to the first two terms of the bound of theorem 3.1: $\mathbf{E}_{h\sim\rho} \mathrm{R}_S^{\ell_{01}}(h)$ is the ρ-average risk over $\mathcal{H}$ on the source sample, and $\mathrm{dis}_\rho(S, T_u)$ measures the ρ-average disagreement between the marginals but is specific to the current model depending on ρ. The other term λ_ρ measures the deviation between the expected joint target and source errors of the individual hypothesis from $\mathcal{H}$ (according to ρ). According to this theory, a good domain adaptation is possible if this deviation is low. However, since we suppose that we do not have any label in the target sample, we cannot control or estimate it. In practice, we suppose that λ_ρ is low and we neglect it. In other words, we assume that the labeling information between the two domains is related and that considering only the marginal agreement and the source labels is sufficient to find a good majority vote. Another important point is that the above theorem improves the one proposed in [GER 13][1]. On the one hand, this bound is not degenerated when the source and target distributions are the same or close. On the other hand, our result contains only the half of $\mathrm{dis}_\rho(S, T)$, contrary to the first bound proposed in [GER 13]. Finally, due to the dependence of $\mathrm{dis}_\rho(S, T)$ and λ_ρ on the learned posterior, our bound is in general incomparable with the ones of theorems 3.1 and 3.4. However, it brings the same underlying idea: supposing that the two domains are sufficiently related, one must look for a model that minimizes a trade-off between its source risk and a distance between the domains' marginal.

6.2.2. *A different philosophy*

Germain *et al.* [GER 16] introduce another domains' divergence to provide an original bound for the PAC–Bayesian setting. Concretely, the authors take advantage of equation [6.2] (recalled below on the target domain $\mathcal{T}$) that expresses the risk of the Gibbs classifier in terms of two terms:

$$\mathop{\mathbf{E}}_{h\sim\rho} \mathrm{R}_{\mathcal{T}}^{\ell_{01}}(h) \;=\; \tfrac{1}{2} d_{\mathcal{T}_\mathbf{X}}(\rho) + e_{\mathcal{T}}(\rho) \qquad [6.4]$$

Put into words, this is half of the expected disagreement that does not rely on labels and the expected joint error. Consequently, one has access to $d_{\mathcal{T}_\mathbf{X}}(\rho)$ even if the target labels are unknown, and the only term that should be upper bounded by terms related to the source domain $\mathcal{S}$ is $e_{\mathcal{T}}(\rho)$. To do this, the authors design a divergence to link $e_{\mathcal{T}}(\rho)$ to $e_{\mathcal{S}}(\rho)$, which is called the β-divergence and defined by

$$\forall q > 0, \quad \beta_q \;=\; \left[\mathop{\mathbf{E}}_{(\mathbf{x},y)\sim\mathcal{S}} \left(\frac{\mathcal{T}(\mathbf{x}, y)}{\mathcal{S}(\mathbf{x}, y)} \right)^q \right]^{\frac{1}{q}}. \qquad [6.5]$$

1 More details are given in the research report [GER 15a] provided by the authors.

The β-divergence can be parameterizable according to the value of $q > 0$, and even allows us to recover well-known distributions' divergence such as the χ^2-distance or the Rényi divergence. Note that the Rényi divergence between source and target domains has led to a non-PAC–Bayesian domain adaptation bound in a multisource setting [MAN 09b]. When $q \to \infty$, we have

$$\beta_\infty = \sup_{(\mathbf{x},y)\in \text{SUPP}(\mathcal{S})} \left(\frac{\mathcal{T}(\mathbf{x},y)}{\mathcal{S}(\mathbf{x},y)} \right), \qquad [6.6]$$

with $\text{SUPP}(\mathcal{S})$ the support of the domain $\mathcal{S}$. This β-divergence gives rise to the following bound.

THEOREM 6.3 ([GER 16]).– *Let $\mathcal{H}$ be a hypothesis space, let $\mathcal{S}$ and $\mathcal{T}$, respectively, be the source and the target domains on $\mathbf{X} \times Y$ and let $q > 0$ be a constant. We have for all posterior distributions ρ on $\mathcal{H}$,*

$$\mathop{\mathbf{E}}_{h\sim\rho} \mathrm{R}_{\mathcal{T}}^{\ell_{01}}(h) \;\leq\; \frac{1}{2} d_{\mathcal{T}_\mathbf{X}}(\rho) + \beta_q \times \Big[e_{\mathcal{S}}(\rho) \Big]^{1-\frac{1}{q}} + \eta_{\mathcal{T}\setminus\mathcal{S}},$$

where

$$\eta_{\mathcal{T}\setminus\mathcal{S}} = \mathop{\mathbf{Pr}}_{(\mathbf{x},y)\sim\mathcal{T}} \Big((\mathbf{x},y) \notin \text{SUPP}(\mathcal{S}) \Big) \sup_{h\in\mathcal{H}} \mathrm{R}_{\mathcal{T}\setminus\mathcal{S}}(h)$$

with $\mathcal{T}\setminus\mathcal{S}$ the distribution of $(\mathbf{x},y)\sim\mathcal{T}$ conditional an $(\mathbf{x},y) \in \text{SUPP}(\mathcal{T})\setminus\text{SUPP}(\mathcal{S})$.

The last term of the bound $\eta_{\mathcal{T}\setminus\mathcal{S}}$, which cannot be estimated without target labels, captures the worst possible risk for the target area not included in $\text{SUPP}(\mathcal{S})$. Note that we have

$$\eta_{\mathcal{T}\setminus\mathcal{S}} \leq \mathop{\mathbf{Pr}}_{(\mathbf{x},y)\sim\mathcal{T}} \big((\mathbf{x},y) \notin \text{SUPP}(\mathcal{S}) \big).$$

An interesting property of theorem 6.3 is that when domain adaptation is not required (*i.e.*, $\mathcal{S} = \mathcal{T}$), the bound is still sound and non-degenerated. Indeed, in this case we have

$$\begin{aligned} \mathrm{R}_{\mathcal{S}}(G_\rho) \;=\; \mathrm{R}_{\mathcal{T}}(G_\rho) \;&\leq\; \tfrac{1}{2} d_{\mathcal{T}_\mathbf{X}}(\rho) + 1 \times [e_{\mathcal{S}}(\rho)]^1 + 0 \\ &=\; \tfrac{1}{2} d_{\mathcal{S}_\mathbf{X}}(\rho) + e_{\mathcal{S}}(\rho) \;=\; \mathrm{R}_{\mathcal{S}}(G_\rho). \end{aligned}$$

Now, we present the PAC–Bayesian generalization bound obtained from the above theorem for which, as in the previous section, a different kind of PAC–Bayesian bound can be derived. The below theorem corresponds to the same approach as the one in the previous section. For the sake of readability, the theorem is expressed when $q \to \infty$.

THEOREM 6.4.– *For any domains $\mathcal{S}$ and $\mathcal{T}$ over $\mathbf{X} \times Y$, any set of voters $\mathcal{H}$, any prior distribution π over $\mathcal{H}$, any $\delta \in (0,1]$, any real numbers $b > 0$ and $c > 0$, with a probability at least $1-\delta$ over the random choices of $S \sim (\mathcal{S})^{m_S}$ and $T_u \sim (\mathcal{T}_\mathbf{X})^{m_T}$, for every posterior distribution ρ on $\mathcal{H}$, we have*

$$\begin{aligned}\mathbf{E}_{h\sim\rho} \mathrm{R}_{\mathcal{T}}^{\ell_{01}}(h) \; &\le c' \, \tfrac{1}{2} \, d_T(\rho) + b' \, e_S(\rho) + \eta_{\mathcal{T}\setminus\mathcal{S}} \\ &\quad + \left(\frac{c'}{m_T \times c} + \frac{b'}{m_S \times b}\right)\left(2 \, \mathrm{KL}(\rho|\pi) + \ln \tfrac{2}{\delta}\right),\end{aligned}$$

where $d_T(\rho)$ and $e_S(\rho)$ are the empirical estimations of the target voters' disagreement and the source joint error, and $b' = \frac{b}{1-e^{-b}} \beta_\infty$ and $c' = \frac{c}{1-e^{-c}}$.

Similarly to the first bound, the above theorem upper bounds the target risk by a trade-off of different terms given by the following atypical quantities:

– the expected disagreement $d_{\mathcal{T}}(\rho)$ that captures second-degree information about the target domain (without any label);

– the divergence between the domains, captured by the β_q-divergence, is not anymore an additive term: it weighs the influence of the expected joint source error $e_S(\rho)$ where parameter q allows one to consider different instances of the β_q-divergence;

– the term $\eta_{\mathcal{T}\setminus\mathcal{S}}$ quantifies the worst feasible target error on the regions where the source domain is not informative for the target task.

6.2.3. *Comparison of the two domain adaptation bounds*

Since the bounds of theorems 6.1 and 6.3 rely on different approximations, the gap between them varies according to the context. As stated above, the main difference lies in the estimable terms. In theorem 6.3, the non-estimable terms are the β-divergence β_q and the term $\eta_{\mathcal{T}\setminus\mathcal{S}}$. Contrary to the non-controllable term λ_ρ of theorem 6.1, these terms do not depend on the *learned* posterior distribution ρ: for every ρ on $\mathcal{H}$, β_q and $\eta_{\mathcal{T}\setminus\mathcal{S}}$ are constant values measuring the relation between the domains for the considered task. Moreover, the fact that the β-divergence is not an additive term but a multiplicative one (as opposed to $\mathrm{dis}_\rho(\mathcal{S}_\mathbf{X}, \mathcal{T}_\mathbf{X}) + \lambda_\rho$ in theorem 6.1) is an important contribution of this new perspective. Consequently, β_q can be viewed as a hyperparameter, allowing us to tune the trade-off between the target voters' disagreement $d_{\mathcal{T}_\mathbf{X}}(\rho)$ and the source joint error $e_{\mathcal{S}}(\rho)$.

Note that, when $e_{\mathcal{T}}(\rho) \ge e_{\mathcal{S}}(\rho)$, we can upper bound the term λ_ρ of theorem 6.1 by using the same trick as in the proof of theorem 6.3. This leads to

$$\begin{aligned} e_{\mathcal{T}}(\rho) \ge e_{\mathcal{S}}(\rho) \quad \Longrightarrow \quad \lambda_\rho &= e_{\mathcal{T}}(\rho) - e_{\mathcal{S}}(\rho) \\ &\le \beta_q \times \left[e_{\mathcal{S}}(\rho)\right]^{1-\frac{1}{q}} + \eta_{\mathcal{T}\setminus\mathcal{S}} - e_{\mathcal{S}}(\rho). \end{aligned}$$

Thus, in this particular case, we can rewrite theorem 6.1's statement for all ρ on $\mathcal{H}$, as

$$\mathop{\mathbf{E}}_{h\sim\rho} \mathrm{R}_{\mathcal{T}}^{\ell_{01}}(h) \leq \mathop{\mathbf{E}}_{h\sim\rho} \mathrm{R}_{\mathcal{S}}^{\ell_{01}}(h) + \frac{1}{2}\,\mathrm{dis}_\rho(\mathcal{S}_{\mathbf{X}}, \mathcal{T}_{\mathbf{X}}) + \beta_q \times \left[e_{\mathcal{S}}(\rho)\right]^{1-\frac{1}{q}} - e_{\mathcal{S}}(\rho) + \eta_{\mathcal{T}\setminus\mathcal{S}}.$$

It turns out that, if $d_{\mathcal{T}_{\mathbf{X}}}(\rho) \geq d_{\mathcal{S}_{\mathbf{X}}}(\rho)$ in addition to $e_{\mathcal{T}}(\rho) \geq e_{\mathcal{S}}(\rho)$, the above statement reduces to that of theorem 6.3. In other words, this occurs in the very particular case where the target disagreement and the target expected joint error are greater than their source counterparts, which may be interpreted as a rather favorable situation. However, theorem 6.3 is tighter in all other cases. This highlights that following the same spirit as the seminal works of Chapter 3 by introducing absolute values in theorem 6.1 proof leads to a crude approximation.

Finally, one of the key points of the generalization bounds of theorems 6.2 and 6.4 is that they suggest algorithms for tackling majority vote learning in the domain adaptation context. Similar to what was done in traditional supervised learning [LAN 02, AMB 06, MCA 11, PAR 12, GER 09], Germain *et al.* [GER 13, GER 16] specialize in these theorems to linear classifiers. In this situation, it is well-known that it makes the risk of a linear classifier and the risk of a (properly parametrized) majority vote coincide, while at the same time it promotes large margin classifiers. We recall these specializations in the following section.

6.3. Specialization to linear classifiers

6.3.1. *Setting and notations*

The approach described below is one of the most popular ones in numerous PAC–Bayesian works (e.g. [LAN 02, AMB 06, MCA 11, PAR 12, GER 09]) that are used to consider PAC–Bayesian framework in the context of linear classification. In order to introduce it, let us consider $\mathcal{H}$ as a set of linear classifiers in a d-dimensional space. Each $h_{\mathbf{w}'} \in \mathcal{H}$ is defined by a weight vector $\mathbf{w}' \in \mathbb{R}^d$ as

$$h_{\mathbf{w}'}(\mathbf{x}) = \mathrm{sign}\left[\mathbf{w}' \cdot \mathbf{x}\right],$$

where $\cdot$ denotes the dot product.

By restricting the prior and the posterior distributions over $\mathcal{H}$ to be Gaussian distributions, Langford *et al.* [LAN 02] have specialized the PAC–Bayesian theory in order to bound the expected risk of any linear classifier $h_{\mathbf{w}} \in \mathcal{H}$. More precisely, given a prior π_0 and a posterior $\rho_{\mathbf{w}}$ defined as spherical Gaussians with an identity covariance matrix, respectively, centered on vectors $\mathbf{0}$ and $\mathbf{w}$, for any $h_{\mathbf{w}'} \in \mathcal{H}$, we have

$$\pi_0(h_{\mathbf{w}'}) = \left(\frac{1}{\sqrt{2\pi}}\right)^d \exp\left(-\frac{1}{2}\|\mathbf{w}'\|^2\right),$$

$$\text{and} \quad \rho_{\mathbf{w}}(h_{\mathbf{w}'}) = \left(\frac{1}{\sqrt{2\pi}}\right)^d \exp\left(-\frac{1}{2}\|\mathbf{w}' - \mathbf{w}\|^2\right).$$

An interesting property of these distributions – also seen as multivariate normal distributions, $\pi_0 = \mathcal{N}(\mathbf{0}, \mathbf{I})$ and $\rho_{\mathbf{w}} = \mathcal{N}(\mathbf{w}, \mathbf{I})$ – is that the prediction of the $\rho_{\mathbf{w}}$-weighted majority vote $B_{\rho_{\mathbf{w}}}$ coincides with the one of the linear classifier $h_{\mathbf{w}}$. Indeed, we have

$$\begin{aligned} \forall \mathbf{x} \in \mathbf{X}, \forall \mathbf{w} \in \mathcal{H}, \quad h_{\mathbf{w}}(\mathbf{x}) &= B_{\rho_{\mathbf{w}}}(\mathbf{x}) \\ &= \text{sign}\left[\mathop{\mathbf{E}}_{h_{\mathbf{w}'} \sim \rho_{\mathbf{w}}} h_{\mathbf{w}'}(\mathbf{x})\right]. \end{aligned}$$

Moreover, the expected risk $\mathbf{E}_{h_{\mathbf{w}'} \sim \rho_{\mathbf{w}}} \mathrm{R}_{\mathcal{D}}^{\ell_{01}}(h_{\mathbf{w}'})$ on a domain $\mathcal{D}$ is then given by[2]

$$\mathop{\mathbf{E}}_{h_{\mathbf{w}'} \sim \rho_{\mathbf{w}}} \mathrm{R}_{\mathcal{D}}^{\ell_{01}}(h_{\mathbf{w}'}) = \mathop{\mathbf{E}}_{(\mathbf{x},y) \sim \mathcal{S}} \Phi_{\mathrm{R}}\left(y \frac{\mathbf{w} \cdot \mathbf{x}}{\|\mathbf{x}\|}\right), \qquad [6.7]$$

where

$$\Phi_{\mathrm{R}}(x) = \tfrac{1}{2}\left[1 - \mathrm{Erf}\left(\tfrac{x}{\sqrt{2}}\right)\right], \qquad [6.8]$$

where $\mathrm{Erf}(\cdot)$ is the Gauss error function defined as

$$\mathrm{Erf}(x) = \frac{2}{\sqrt{\pi}} \int_0^x \exp\left(-t^2\right) dt. \qquad [6.9]$$

Here, $\Phi_{\mathrm{R}}(x)$ can be seen as a smooth surrogate of the $0-1$ loss function $\mathbf{I}[x \leq 0]$ relying on $y \frac{\mathbf{w} \cdot \mathbf{x}}{\|\mathbf{x}\|}$. This function Φ_{R} is sometimes called the probit-loss (e.g., [MCA 11]). It is worth noting that $\|\mathbf{w}\|$ plays an important role in the definition of $\mathbf{E}_{h_{\mathbf{w}'} \sim \rho_{\mathbf{w}}} \mathrm{R}_{\mathcal{D}}(h_{\mathbf{w}'})$, but not in that of $\mathrm{R}_{\mathcal{D}}^{\ell_{01}}(h_{\mathbf{w}})$. Indeed, $\mathbf{E}_{h_{\mathbf{w}'} \sim \rho_{\mathbf{w}}} \mathrm{R}_{\mathcal{D}}(h_{\mathbf{w}'})$ tends to $\mathrm{R}_{\mathcal{D}}^{\ell_{01}}(h_{\mathbf{w}})$ as $\|\mathbf{w}\|$ grows, which can provide very tight bounds (see the empirical analyses of [AMB 06, GER 09]). Finally, the KL-divergence between $\rho_{\mathbf{w}}$ and π_0 becomes simply

$$\begin{aligned} \mathrm{KL}(\rho_{\mathbf{w}} | \pi_0) &= \mathrm{KL}(\mathcal{N}(\mathbf{w}, \mathbf{I}) | \mathcal{N}(\mathbf{0}, \mathbf{I})) \\ &= \tfrac{1}{2}\|\mathbf{w}\|^2, \end{aligned}$$

and turns out to be a measure of complexity of the learned classifier.

6.3.2. *Domain disagreement, expected disagreement and joint error of linear classifiers*

The authors express the domain disagreement $\mathrm{dis}_{\rho_{\mathbf{w}}}(\mathcal{S}_{\mathbf{X}}, \mathcal{T}_{\mathbf{X}})$, the expected disagreement $d_{\mathcal{D}_{\mathbf{X}}}(\rho_{\mathbf{w}})$ and the expected joint error $e_{\mathcal{D}}(\rho_{\mathbf{w}})$ in terms of linear classifiers as follows.

2 The calculations leading to equation [6.7] can be found in [LAN 05].

First, for any marginal $\mathcal{D}_{\mathbf{X}}$, the expected disagreement for linear classifiers is:

$$\begin{aligned} d_{\mathcal{D}_{\mathbf{X}}}(\rho_{\mathbf{w}}) &= \mathop{\mathbf{E}}_{\mathbf{x}\sim\mathcal{D}_{\mathbf{X}}} \mathop{\mathbf{E}}_{(h,h')\sim\rho_{\mathbf{w}}^2} \ell_{01}\big(h(\mathbf{x}), h'(\mathbf{x})\big) \\ &= \mathop{\mathbf{E}}_{\mathbf{x}\sim\mathcal{D}_{\mathbf{X}}} \mathop{\mathbf{E}}_{(h,h')\sim\rho_{\mathbf{w}}^2} \mathbf{I}\left[h(\mathbf{x}) \neq h'(\mathbf{x})\right] \\ &= 2 \mathop{\mathbf{E}}_{\mathbf{x}\sim\mathcal{D}_{\mathbf{X}}} \mathop{\mathbf{E}}_{(h,h')\sim\rho_{\mathbf{w}}^2} \mathbf{I}\left[h(\mathbf{x}) = 1\right] \mathbf{I}\left[h'(\mathbf{x}) = -1\right] \\ &= 2 \mathop{\mathbf{E}}_{\mathbf{x}\sim\mathcal{D}_{\mathbf{X}}} \mathop{\mathbf{E}}_{h\sim\rho_{\mathbf{w}}} \mathbf{I}\left[h(\mathbf{x}) = 1\right] \mathop{\mathbf{E}}_{h'\sim\rho_{\mathbf{w}}} \mathbf{I}\left[h'(\mathbf{x}) = -1\right] \\ &= 2 \mathop{\mathbf{E}}_{\mathbf{x}\sim\mathcal{D}_{\mathbf{X}}} \Phi_{\mathrm{R}}\left(\frac{\mathbf{w}\cdot\mathbf{x}}{\|\mathbf{x}\|}\right) \Phi_{\mathrm{R}}\left(-\frac{\mathbf{w}\cdot\mathbf{x}}{\|\mathbf{x}\|}\right) \\ &= \mathop{\mathbf{E}}_{\mathbf{x}\sim\mathcal{D}_{\mathbf{X}}} \Phi_d\left(\frac{\mathbf{w}\cdot\mathbf{x}}{\|\mathbf{x}\|}\right), \end{aligned} \qquad [6.10]$$

where

$$\Phi_d(x) \;=\; 2\,\Phi_{\mathrm{R}}(x)\,\Phi_{\mathrm{R}}(-x). \qquad [6.11]$$

Thus, the domain disagreement for linear classifiers is

$$\begin{aligned} \mathrm{dis}_{\rho_{\mathbf{w}}}(\mathcal{S}_{\mathbf{X}}, \mathcal{T}_{\mathbf{X}}) &= \left| d_{\mathcal{S}_{\mathbf{X}}}(\rho_{\mathbf{w}}) - d_{\mathcal{T}_{\mathbf{X}}}(\rho_{\mathbf{w}}) \right| \\ &= \left| \mathop{\mathbf{E}}_{\mathbf{x}\sim\mathcal{S}_{\mathbf{X}}} \Phi_d\left(\frac{\mathbf{w}\cdot\mathbf{x}}{\|\mathbf{x}\|}\right) - \mathop{\mathbf{E}}_{\mathbf{x}\sim\mathcal{T}_{\mathbf{X}}} \Phi_d\left(\frac{\mathbf{w}\cdot\mathbf{x}}{\|\mathbf{x}\|}\right) \right|. \end{aligned} \qquad [6.12]$$

Following a similar approach, the expected joint error is, for all $\mathbf{w} \in \mathbb{R}^d$,

$$\begin{aligned} e_{\mathcal{D}}(\rho_{\mathbf{w}}) &= esp(\mathbf{x}, y) \sim \mathcal{D} \mathop{\mathbf{E}}_{h\sim\rho_{\mathbf{w}}} \mathop{\mathbf{E}}_{h'\sim\rho_{\mathbf{w}}} \ell_{01}\left(h(\mathbf{x}), y\right) \times \ell_{01}\left(h'(\mathbf{x}), y\right) \\ &= \mathop{\mathbf{E}}_{(\mathbf{x},y)\sim\mathcal{D}} \mathop{\mathbf{E}}_{h\sim\rho_{\mathbf{w}}} \ell_{01}\left(h(\mathbf{x}), y\right) \mathop{\mathbf{E}}_{h'\sim\rho_{\mathbf{w}}} \ell_{01}\left(h'(\mathbf{x}), y\right) \\ &= \mathop{\mathbf{E}}_{(\mathbf{x},y)\sim\mathcal{P}} \Phi_e\left(y\,\frac{\mathbf{w}\cdot\mathbf{x}}{\|\mathbf{x}\|}\right), \end{aligned} \qquad [6.13]$$

with

$$\Phi_e(x) \;=\; \left[\Phi_{\mathrm{R}}(x)\right]^2. \qquad [6.14]$$

Functions Φ_e and Φ_d defined above can be interpreted as loss functions for linear classifiers (illustrated by Figure 6.1).

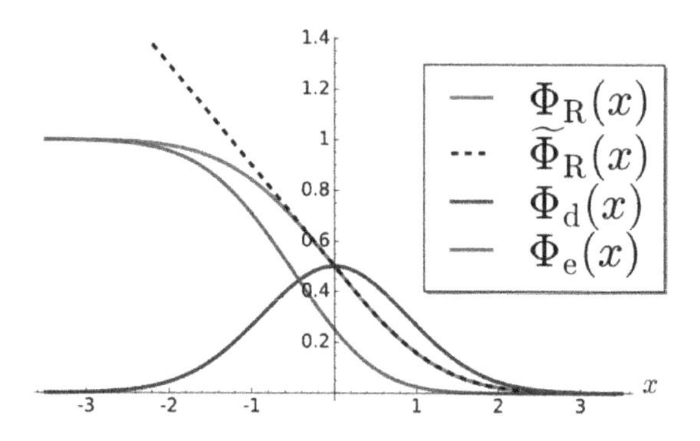

Figure 6.1. *Loss functions given by the specialization to linear classifiers (figure from [GER 16])*

6.3.3. *Domain adaptation bounds for linear classifiers*

Theorems 6.1 and 6.3 (with $q \to \infty$) specialized to linear classifiers give the two following corollaries. Note that, as mentioned before, we have

$$\begin{aligned} \mathrm{R}_{\mathcal{D}}^{\ell_{01}}(h_{\mathbf{w}}) &= \mathrm{R}_{\mathcal{D}}^{\ell_{01}}(B_{\rho_{\mathbf{w}}}) \\ &\leq 2 \mathop{\mathbf{E}}_{h_{\mathbf{w}'} \sim \rho_{\mathbf{w}}} \mathrm{R}_{\mathcal{D}}^{\ell_{01}}(h_{\mathbf{w}'}). \end{aligned}$$

COROLLARY 6.1.– *Let $\mathcal{S}$ and $\mathcal{T}$, respectively, be the source and the target domains on $\mathbf{X} \times Y$. For all $\mathbf{w} \in \mathbb{R}^d$, we have*

$$\begin{aligned} \mathrm{R}_{\mathcal{T}}^{\ell_{01}}(h_{\mathbf{w}}) &\leq 2\,\mathrm{R}_{\mathcal{S}}^{\ell_{01}}(h_{\mathbf{w}}) + \mathrm{dis}_{\rho_{\mathbf{w}}}(\mathcal{S}_{\mathbf{X}}, \mathcal{T}_{\mathbf{X}}) + 2\lambda_{\rho_{\mathbf{w}}} \\ &= 2\,\mathrm{R}_{\mathcal{S}}^{\ell_{01}}(h_{\mathbf{w}}) + \left| \mathop{\mathbf{E}}_{\mathbf{x} \sim \mathcal{S}_{\mathbf{X}}} \Phi_d\left(\frac{\mathbf{w} \cdot \mathbf{x}}{\|\mathbf{x}\|}\right) - \mathop{\mathbf{E}}_{\mathbf{x} \sim \mathcal{T}_{\mathbf{X}}} \Phi_d\left(\frac{\mathbf{w} \cdot \mathbf{x}}{\|\mathbf{x}\|}\right) \right| + 2\lambda_{\rho_{\mathbf{w}}}. \end{aligned}$$

COROLLARY 6.2.– *Let $\mathcal{S}$ and $\mathcal{T}$, respectively, be the source and the target domains on $\mathbf{X} \times Y$. For all $\mathbf{w} \in \mathbb{R}^d$, we have*

$$\begin{aligned} \mathrm{R}_{\mathcal{T}}^{\ell_{01}}(h_{\mathbf{w}}) &\leq d_{\mathcal{T}_{\mathbf{X}}}(\rho_{\mathbf{w}}) + 2\,\beta_\infty \times e_{\mathcal{S}}(\rho_{\mathbf{w}}) + 2\,\eta_{\mathcal{T}\setminus\mathcal{S}} \\ &= \mathop{\mathbf{E}}_{\mathbf{x} \sim \mathcal{D}_{\mathbf{X}}} \Phi_d\left(\frac{\mathbf{w} \cdot \mathbf{x}}{\|\mathbf{x}\|}\right) + 2\,\beta_\infty \mathop{\mathbf{E}}_{(\mathbf{x},y) \sim \mathcal{P}} \Phi_e\left(y\,\frac{\mathbf{w} \cdot \mathbf{x}}{\|\mathbf{x}\|}\right) + 2\,\eta_{\mathcal{T}\setminus\mathcal{S}}. \end{aligned}$$

For fixed values of β_∞ and $\eta_{\mathcal{T}\setminus\mathcal{S}}$, the target risk $\mathrm{R}_{\mathcal{T}}^{\ell_{01}}(h_{\mathbf{w}})$ is upper bounded by a (β_∞-weighted) sum of two losses. The expected Φ_e-loss (i.e. the joint error) is computed on the (labeled) source domain; it aims to label the source examples correctly, but is more permissive on the required margin than the Φ-loss. The expected Φ_d-loss (i.e. the disagreement) is computed on the target (unlabeled) domain; it promotes large *unsigned* target margins. Hence, corollary 6.2 reflects that some source errors may be allowed if the separation of the target domain is improved. Figure 6.1 leads to an insightful geometric interpretation of the two domain adaptation trade-offs promoted by corollaries 6.1 and 6.2.

6.3.4. *Generalization bounds*

For the sake of completeness, we provide the PAC–Bayesian generalization bounds associated with Corollaries 6.1 and 6.2.

COROLLARY 6.3.– *For any domains $\mathcal{S}$ and $\mathcal{T}$ over $\mathbf{X} \times Y$, any $\delta \in (0,1]$, any $\omega > 0$ and $a > 0$, with a probability of at least $1 - \delta$ over the choices of $S \sim (\mathcal{S})^m$ and $T_u \sim (\mathcal{T}_{\mathbf{X}})^m$, we have, for all $\mathbf{w} \in \mathbb{R}^d$,*

$$\begin{aligned} \mathrm{R}_{\mathcal{T}}^{\ell_{01}}(h_{\mathbf{w}}) \;&\leq 2\,\omega'\,\mathrm{R}_{S}^{\ell_{01}}(h_{\mathbf{w}}) + a'\,\mathrm{dis}_{\rho_{\mathbf{w}}}(S,T_u) + 2\lambda_{\rho_{\mathbf{w}}} \\ &\quad + 2\left(\frac{\omega'}{\omega} + \frac{a'}{a}\right)\frac{\|\mathbf{w}\|^2 + \ln\frac{3}{\delta}}{m} + (a'-1), \end{aligned}$$

where $\mathrm{R}_{S}^{\ell_{01}}(h_{\mathbf{w}})$ and $\mathrm{dis}_{\rho_{\mathbf{w}}}(S,T_u)$ are the empirical estimates of the target risk and the domain disagreement; $\lambda_{\rho_{\mathbf{w}}}$ is obtained using equation [6.3], $\omega' = \frac{\omega}{1-e^{-\omega}}$ and $a' = \frac{2a}{1-e^{-2a}}$.

COROLLARY 6.4.– *For any domains $\mathcal{S}$ and $\mathcal{T}$ over $\mathbf{X} \times Y$, any $\delta \in (0,1]$, any $\omega > 0$ and $a > 0$, with a probability at least $1 - \delta$ over the choices of $S \sim (\mathcal{S})^m$ and $T_u \sim (\mathcal{T}_{\mathbf{X}})^m$, we have, for all $\mathbf{w} \in \mathbb{R}^d$,*

$$\begin{aligned} \mathrm{R}_{\mathcal{T}}^{\ell_{01}}(h_{\mathbf{w}}) \;&\leq c'\,d_T(\rho_{\mathbf{w}}) + 2\,b'\,e_S(\rho_{\mathbf{w}}) \\ &\quad + 2\,\eta_{\mathcal{T}\setminus\mathcal{S}} + 2\left(\tfrac{c'}{m_T \times c} + \tfrac{b'}{m_S \times b}\right)\left(\|\mathbf{w}\|^2 + \ln\tfrac{2}{\delta}\right), \end{aligned}$$

where $d_T(\rho_{\mathbf{w}})$ and $e_S(\rho_{\mathbf{w}})$ are the empirical estimations of the target voters' disagreement and the source joint error, and $b' = \frac{b}{1-e^{-b}}\,\beta_\infty$ and $c' = \frac{c}{1-e^{-c}}$.

The bounds of corollaries 6.3 and 6.4 allowed the authors to derive two domain adaptation algorithms specialized to linear classifiers, respectively, called PBDA and DALC by minimizing the trade-off involved in the bounds (see [GER 13, GER 16] for more details on the derivation and the evaluation of these algorithms).

Note that McNamara and Balcan [MCN 17] made use of the PAC–Bayesian framework to derive a generalization bound in a domain adaptation related framework that is fine tuning in deep learning. This setting corresponds to a scenario where one wants to adapt a network trained for a given domain to a similar one. The authors obtained a bound that does not directly involve the notion of divergence between the domains, but a function that measures a transferability property between the two domains.

6.4. Summary

In this chapter, we recalled the two domain adaptation analyses for the PAC–Bayesian framework proved by Germain *et al.* [GER 13, GER 16] that focuses on models taking the form of a majority vote over a set of classifiers. The first analysis is based on the well-established principle of the seminal results of domain adaptation, while the second one brings a novel perspective on domain adaptation.

More precisely, the first result of this chapter follows the underlying philosophy of the seminal works of Ben-David *et al.* and Mansour *et al.* of Chapter 3. In other words, the authors derived an upper bound on the target risk because of the domains' divergence measure suitable for the PAC–Bayesian setting. This divergence is expressed as the average deviation between the disagreement over a set of classifiers on the source and target domains, while both $\mathcal{H}\Delta\mathcal{H}$-divergence and discrepancy are defined in terms of the worst-case deviation. This leads to a bound that takes the form of a trade-off between the source risk, the domains' divergence and a term that captures the ability to adapt for the current task. Then, we recalled another domain adaptation bound that takes advantage of the inherent behavior of the target risk in the PAC–Bayesian setting. The upper obtained bound is different from the original one as it expresses a trade-off between the disagreement on the target domain only, the joint errors of the classifiers on the source domain only and a term reflecting the worst case error in regions where the source domain is non-informative. Contrary to the first bound and those of the previous chapters, the divergence is not an additive term but is a factor weighing the importance of the source information.

These analyses were combined with PAC–Bayesian generalization bounds of Chapter 1 and involved an additional term that measures the deviation of the learned majority vote to the *a priori* knowledge we have on the majority vote. Additionally, as done in many classical PAC–Bayesian analyses, the bounds were specialized to linear classifiers to facilitate the deviation of domain adaptation algorithms[3].

3 Since this book is mainly dedicated to theoretical analyses of domain adaptation, we did not present the corresponding algorithms. The reader can refer to the original papers [GER 13, GER 16] for more details.

7

Domain Adaptation Theory Based on Algorithmic Properties

7.1. Introduction

In this chapter, we first review the work of Mansour and Schain [MAN 14], where they derive generalization bound based on the algorithm with regard to the algorithmic robustness of Xu and Mannor [XU 10] recalled in Chapter 1.

Then we review the works of Kuzborskij and Orabona [KUZ 13] and Perrot and Habrard [PER 15b] based on the algorithmic stability recalled in Chapter 1. Note that these domain adaptation contributions are based on a slightly different setting where one does not have access to the source examples, but rather has access to hypotheses learned from the source domain.

7.2. Robust domain adaptation

7.2.1. λ*-shift a measure of similarity between domains*

Based on the notion of algorithmic robustness [XU 10] (recalled in section 1.2.5), Mansour and Schain [MAN 14] define the λ-shift that encodes a prior knowledge on the deviation between source and target domain. The objective is to guarantee that the loss of the algorithm on source and target domains is similar in regions of $\mathbf{X} \times Y$. Since one does not necessarily have access to target labels, the authors propose to consider the conditional distribution of the label in a given region and the relation to its sampled value over the given labeled sample S. Let ρ be a distribution over the label space Y. The probability of a given label $y \in Y$ is denoted by σ^y, and the total probability of the other labels is $\sigma^{-y} = 1 - \sigma^y$.

DEFINITION 7.1 ([MAN 14]).– *Let σ and ρ be two distribution's over Y. ρ is λ-shift with respect to σ, denoted $\rho \in \lambda(\sigma)$, if for all $y \in Y$ we have $\rho^y \leq \sigma^y + \lambda\sigma^{-y}$, and $\rho^y \geq \sigma^y(1-\lambda)$. If for some $y \in Y$, we have $\rho^y = \sigma^y + \lambda\sigma^{-y}$ we say that ρ is strict-λ-shift with respect to σ.*

Note that, for sake of simplicity, for $\rho \in \lambda(\sigma)$ the upper bound and the lower bound of the probability ρ^y are, respectively, denoted by:

$$\bar{\lambda}^y(\sigma) = \sigma^y + \lambda(1-\sigma^y), \qquad \text{and} \qquad \underline{\lambda}^y(\sigma) = \sigma^y(1-\lambda).$$

The above definition means that assuming the λ-shift between two distributions on Y implies a restriction on the deviation between the probability of a label on the distributions, this shift may be at most a λ portion of the probability of the other labels or of the probability of the label. Note that $\lambda = 1$ and $\lambda = 0$ correspond to the no restriction case and the total restriction case.

7.2.2. *Domain adaptation bounds using robustness*

To deal with domain adaptation, the authors assume that $\mathbf{X} \times Y$ can be partitioned into M disjoint subsets defined as: $\mathbf{X} \times Y = \cup_{i,j} \mathbf{X}_i \times Y_j$, where the input space is partitioned as $\mathbf{X} = \cup_{i=1}^{M_\mathbf{X}}$ and the output space as $Y = \cup_{j=1}^{M_Y} Y_j$, and $M = M_\mathbf{X} M_Y$. Note that, given a (M,ϵ)-robust algorithm, the hypothesis learned has then an ϵ variation in the loss of each region $\mathbf{X}_i \times Y_j$.

THEOREM 7.1 ([MAN 14]).– *Let $\mathcal{A}$ be a (M,ϵ)-robust algorithm with respect to a loss function $\ell : \mathbf{X} \times Y$ such that $0 \leq \ell(h(\mathbf{x}, y) \leq M_l$, for all $(\mathbf{x}, y) \in (\mathbf{X} \times Y)$ and $h \in \mathcal{H}$. If $\mathcal{S}$ is λ-shift of $\mathcal{T}$ with respect to the partition of $\mathbf{X}$ for any $\delta \in (0,1]$, the following bound holds with probability at least $1-\delta$, over the random draw of the sample S from $\mathcal{S}$, and of the sample T from $\mathcal{T}$ of size m,*

$$\forall h \in \mathcal{H},\ \mathrm{R}^\ell_\mathcal{T}(h) \leq \sum_{i=1}^{M_\mathbf{x}} T(\mathbf{X}_i)\ell^\lambda_\mathcal{S}(h,\mathbf{X}_i) + \epsilon + M_\ell\sqrt{\frac{2M\ln 2 + 2\ln\frac{1}{\delta}}{m}},$$

where $T(\mathbf{X}_i) = \frac{1}{m}\left|\{\mathbf{x} \in T \cap \mathbf{X}_i\}\right|$ is the ratio of target points in the region $\mathbf{X}_i$, and

$$\forall i \in \{1,\ldots,M_\mathbf{X}\},\quad \ell^\lambda_\mathcal{S}(h,\mathbf{X}_i) \leq \max_{y\in Y}\left\{\ell_i(h,y)\bar{\lambda}^y(\mathcal{S}_i) + \sum_{y'\neq y}\ell_i(h,y')\underline{\lambda}^{y'}(\mathcal{S}_i)\right\},$$

with

$$\ell_i(h,y) = \begin{cases} \max_{\mathbf{x}\in\mathbf{X}_i\times y}\ell(h(\mathbf{x}),y) & \text{if } S\cap\mathbf{X}_i\times y \neq \emptyset \\ M_\ell & \text{otherwise.}\end{cases}$$

The main difference between this domain adaptation result and the original robustness bound of theorem 1.8 of Chapter 1 stands in the first term. In the latter one, that is an upper bound on the source risk, the first term $\frac{1}{|S|}\sum_{(x,y)\in S}\ell_S(h(x),y)$ simply corresponds to the empirical error of the model learned on the source sample. In the former bound, that upper bounds the target risk, while the first term $\sum_{i=1}^{M_{\mathbf{x}}} T(\mathbf{X}_i)\ell_S^{\lambda}(h,\mathbf{X}_i)$ depends also on the empirical risk on the source sample, which is a combination of the λ-shifted source risk of each region weighted by the ratio of target points in the region.

Note that another domain adaptation analysis with algorithmic robustness has also been proposed in [MOR 12b], but in particular context based on similarity functions.

7.3. Hypothesis transfer learning

In this section, we review theoretical results for a slightly different setting called *hypothesis transfer learning (HTL)*. It differs from the classic domain adaptation framework by the fact that only source hypotheses are available from the source domain but not the source (labeled) data, and one generally assumes to have access to only a *small* training sample from the target domain. As a result, HTL does not require an explicit access to the source domain or any assumption about the relatedness of the source and target distributions. Another advantage is to avoid the use of costly procedures for the estimation of adaptation parameters and the need of storing abundant source data. These advantages have inspired many approaches that demonstrated the interest of HTL in practice [FEI 06, YAN 07, TOM 10, MAN 08].

As reviewed earlier in this manuscript, theoretical results in domain adaptation take the form of learning bounds on the target risk that depend on the source risk, a divergence term between the source and target distribution, and a notion of minimal joint error. The divergence term in particular is necessary unless an algorithm has access to some labeled training data from the target domain [BEN 10b]. In contrast, HTL methods depend on the way the source hypothesis has been generated, which is not covered by domain adaptation theoretical frameworks. The fact that the source domain is not accessible makes the domain adaptation bounds unadapted to HTL.

In this part, first a general definition for the HTL problem is provided. Then, we present results for HTL in context of regularized least squares (RLS) problems. This result is then extended by a general ERM strategy for transferring multiple auxiliary hypothesis. In a second part, we also provide some similar results obtained in the context of metric learning.

7.3.1. *Problem set-up*

As mentioned above, until now we have considered that we have access to a labeled source sample S. In HTL, we do not have access to such a set. Instead we have a hypothesis learned from the source sample S and the hypothesis space $\mathcal{H}_{\mathcal{S}}$ associated with the source domain $\mathcal{S}$. The objective is thus to use the hypothesis $h_{\mathcal{S}} \in \mathcal{H}_{\mathcal{S}}$ and (labeled) target sample $T = \{(\mathbf{x}_i, y_i)\}_{i=1}^m \sim (\mathcal{T})^m$ for learning a target model better than the one we could learn from T only. An HTL algorithm $\mathcal{A}$ is then defined as

$$\mathcal{A} : (\mathbf{X} \times Y)^m \times \mathcal{H}_{\mathcal{S}} \mapsto \mathcal{H},$$

where $\mathcal{A}$ maps any (labeled) target sample $T \sim (\mathcal{T})^m$ and a source hypothesis $h_{\mathcal{S}} \in \mathcal{H}_{\mathcal{S}}$ onto a target hypothesis $h \in \mathcal{H}$.

The objective of an HTL problem can be formalized by the following definition.

DEFINITION 7.2 (Usefulness and Collaboration [KUZ 18]).– *An hypothesis $h_{src} \in \mathcal{H}_{\mathcal{S}}$ is useful for $\mathcal{A}$ with respect to distribution $\mathcal{S}$ and a training sample S of size m if*

$$\mathop{\mathbf{E}}_{S \sim (\mathcal{S})^m} [\mathrm{R}_{\mathcal{D}}(\mathcal{A}(S, h_{src}))] < \mathop{\mathbf{E}}_{S \sim (\mathcal{S})^m} [\mathrm{R}_{\mathcal{D}}(\mathcal{A}(S, \mathbf{0}))].$$

A hypothesis $h_{src} \in \mathcal{H}_{\mathcal{S}}$ and a distribution $\mathcal{D}$ collaborate [BEN 13] for $\mathcal{A}$, with respect to a training sample S of size m, if

$$\mathop{\mathbf{E}}_{S \sim (\mathcal{S})^m} [\mathrm{R}_{\mathcal{S}}(\mathcal{A}(S, h_{src}))] < \min \left\{ \mathrm{R}_{\mathcal{S}}(\mathcal{A}(\emptyset, h_{src}), \mathop{\mathbf{E}}_{S \sim (\mathcal{S})^m} [\mathrm{R}_{\mathcal{D}}(\mathcal{A}(S, \mathbf{0}))] \right\}.$$

This definition provides two interesting properties for an HTL algorithm. The notion of usefulness corresponds to the case where the algorithm $\mathcal{A}$ allows one to infer a model with a lower risk by using the source hypothesis. The collaboration refers to the case where the access to both the source hypothesis h_{src} and the sample S used together helps to increase performance in comparison to the case where they are used separately. If any one of these two properties is not satisfied, then the resulting learning procedure leads to a situation called *negative transfer*. As a result, what really matters here is the capacity to observe a decrease in the generalization error on the target domain.

7.3.2. *A biased regularized least squares algorithm for HTL*

We first begin by a quick recap of the classic RLS algorithm. Given a (labeled) learning sample $T = \{(\mathbf{x}_i, y_i)\}_{i=1}^m \sim (\mathcal{T})^m$ with the assumptions that $y_i \in [-B, B]$ with $B \in \mathbb{R}$ and $\mathbf{x}_i \in \mathbf{R}^d$ with $\|\mathbf{x}\| \leq 1$. RLS method aims at solving the following optimization problem:

$$\min_{\mathbf{w} \in \mathbf{R}^d} \left\{ \frac{1}{m} \sum_{i=1}^m (\mathbf{w}^T \mathbf{x}_i - y_i)^2 + \lambda \|\mathbf{w}\|^2 \right\}.$$

RLS has nice theoretical properties and its solution can be expressed in closed form.

Now, we consider a source hypothesis of the form $h_{\text{src}}(\mathbf{x}) = \mathbf{x}^T\mathbf{w}_0$, where $\mathbf{w}_0$ corresponds to the parameters of the source hypothesis that are supposed to be from the same space as $\mathbf{w}$. In [ORA 09], it has been proposed to use a biased regularization with respect to $\mathbf{w}_0$ as

$$\min_{\mathbf{w}\in\mathbf{R}^d}\left\{\frac{1}{m}\sum_{i=1}^{m}(\mathbf{w}^T\mathbf{x}_i - y_i)^2 + \lambda\|\mathbf{w}-\mathbf{w}_0\|^2\right\}.$$

In the formulation above, we can clearly see that the source hypothesis represented by $\mathbf{w}_0$ acts as a bias that tends to make the learned model close to $\mathbf{w}_0$ if the learning sample is compatible with it.

Following the result of [KUZ 13], we present a more general version where the target hypothesis to learn is defined by

$$h_T(\mathbf{x}) = \text{tr}_C\left(\mathbf{x}^\top\hat{\mathbf{w}}_T\right) + h_{\text{src}}(\mathbf{x}), \tag{7.1}$$

where

$$\hat{\mathbf{w}}_T = \underset{\mathbf{w}}{\text{argmin}}\ \frac{1}{m}\sum_{i=1}^{m}\left(\mathbf{w}^\top\mathbf{x}_i - y_i + h_{\text{src}}(\mathbf{x}_i)\right)^2 + \lambda\|\mathbf{w}\|^2,$$

and the truncation function $\text{tr}_C(a)$ is defined as

$$\text{tr}_C(a) = \min\left[\max\left(a, -C\right), C\right].$$

This version has two main advantages. First, it can be seen as a generalization of usual biased RLS algorithm by allowing to consider any type of source model h_{src}. From this version of biased RLS algorithm, we can retrieve the usual formulation when $C = \infty$ and $h_{\text{src}}(\mathbf{x}) = \mathbf{x}^\top\mathbf{w}_0$, where $\mathbf{w}_0$ and $\mathbf{w}_T$ belongs to the same space.

From the theoretical standpoint, the objective is to bound the expected risk associated with this algorithm in terms of the characteristics of the source model h_{src}. The proposed result is based upon the *leave-one-out (LOO)* risk over a sample T defined as

$$\text{R}^{\text{loo}}_{\hat{T}}(\mathcal{A}, T) = \frac{1}{m}\sum_{i=1}^{m}\ell(\mathcal{A}_{T^{\setminus i}}, (\mathbf{x}_i, y_i)),$$

where $\mathcal{A}_{T^{\setminus i}}$ represents the model learned by algorithm $\mathcal{A}$ from the sample T without example $(\mathbf{x}_i, y_i)$. The first result related to HTL can be now presented in the following theorem.

THEOREM 7.2 ([KUZ 13]).– *Set* $\lambda \geq \frac{1}{m}$. *If* $C \geq B + \|h_S\|_\infty$, *then for any hypothesis learned by the algorithm presented in equation* [7.1], *with probability at least* $1 - \delta$ *over any sample* T *of size* m *i.i.d. from* $\mathcal{T}$, *we have*

$$\mathrm{R}_{\mathcal{T}}(h_T) - \mathrm{R}_{\hat{\mathcal{T}}}^{loo}(h_T, T) = \mathcal{O}\left(C \frac{\sqrt[4]{\mathrm{R}_{\mathcal{T}}(h_{src})\mathrm{tr}_{C^2}\left(\frac{\mathrm{R}_{\mathcal{T}}(h_{src})}{\lambda}\right) + \mathrm{R}_{\mathcal{T}}^2(h_{src})}}{\sqrt{m}\delta\lambda^{3/4}} \right).$$

If $C = \infty$, *then we have*

$$\mathrm{R}_{\mathcal{T}}(h_T) - \mathrm{R}_{\hat{\mathcal{T}}}^{loo}(h_T, T) = \mathcal{O}\left(\frac{\sqrt{\mathrm{R}_{\mathcal{T}}(h_{src})}(\|h_{src}\|_\infty + B)}{\sqrt{m}\delta\lambda} \right).$$

According to [KUZ 18], we can draw the following implications:

1) If one considers the null hypothesis as a source model, *i.e.*, $h_{\text{src}} = \mathbf{0}$, then we fall in a classic learning setting without any source hypothesis and thus without any transfer learning. When $C = \infty$, the generalization bound is bounded by $\mathcal{O}\left(\frac{B}{\sqrt{m}\lambda}\right)$, which is similar to the results obtained for classic RLS algorithms [BOU 02], but the bound can be improved by setting C with respect to the range $[-B; B]$;

2) If one considers a source hypothesis different from the null hypothesis, the quantity $\mathrm{R}_{\mathcal{T}}(h_{\text{src}})$ is crucial to measure the influence of the source hypothesis. Making the parallel with domain adaptation, it can be interpreted as having a similar information as the one given by the divergence between the source and target distributions as discussed in Chapter 3. Indeed, it provides a direct information on the quality of the source hypothesis with respect to the target domain. This term is multiplicative while divergence-based domain adaptation results usually include the divergence as an additive term, even though the two types of results are not comparable in general. Additionally, if we have that $\frac{\mathrm{R}_{\mathcal{T}}(h_{\text{src}})}{\lambda}$ tends to zero, then $\mathrm{R}_{\mathcal{T}}(h_T) - \mathrm{R}_{\hat{\mathcal{T}}}^{\text{loo}}(h_T, T)$ tends to zero implying that the target true risk converges to the LOO risk. Moreover, when $\frac{\mathrm{R}_{\mathcal{T}}(h_{\text{src}})}{\lambda}$ is small enough with respect to λ, then the bound remains informative even for small training sets. This means that when the source hypothesis is good enough on the target domain, then transfer learning helps to learn better hypothesis on the target domain even with small training samples. Considering the algorithmic stability framework used by the authors to prove this theorem, we note that when the source hypothesis is better on the target domain, the HTL algorithm has better stability and thus a better generalization guarantee on the target domain. The good stability property is ensured by the fact that the hypothesis generated for the target domain remains not too far from the source one;

3) If $\frac{\mathrm{R}(h_{\text{src}})}{\lambda}$ is high, then one needs more target labeled data to provide a more reliable hypothesis on the target. The domains are then not related, and the source

hypothesis does not bring any useful information. Actually, the convergence rate remains the same as in a non-transfer setting. This approach is thus in some sense robust to negative transfer, which is a desirable property;

4) If the source hypothesis can be cast as a weighted combination of n source hypotheses, that is $h_{\text{src}} = \sum_{i=1}^{n} \beta_i h_{\text{src}}^i$, then the weights can be tuned from the target sample in a more accurate way than similar multisource strategies used in multisource domain adaptation. Indeed, we can have the following simplified interpretation [KUZ 13, KUZ 18]

$$\mathrm{R}_{\mathcal{T}}(h_T) \leq \min_{\boldsymbol{\beta} \in \mathbb{R}^n} \mathrm{R}_{\hat{T}}^{\text{loo}}(h_T, T) + \mathcal{O}\left(\frac{\|\boldsymbol{\beta}\|}{\sqrt{m}\lambda^{3/4}}\right).$$

The last inequality implies that is it reasonable to minimize the empirical LOO risk with respect $\boldsymbol{\beta}$ under the hypothesis to be able to control and constrain $\|\boldsymbol{\beta}\|$.

7.3.3. *HTL for regularized ERM with smooth and convex loss functions*

In this section, we consider the setting of [KUZ 17] where the source hypothesis is expressed as a weighted combination of different source hypotheses

$$h_{\text{src}}^{\boldsymbol{\beta}}(\mathbf{x}) = \sum_{i=1}^{n} \beta_i h_{\text{src}}^i(\mathbf{x}),$$

and where the target hypothesis is defined as

$$h_{\boldsymbol{w},\boldsymbol{\beta}}(\mathbf{x}) = \langle \boldsymbol{w}, \mathbf{x} \rangle + h_{\text{src}}^{\boldsymbol{\beta}}(\mathbf{x}).$$

The relevance of the different source hypotheses is then characterized by their associated weight given by the vector $\boldsymbol{\beta}$.

The transfer learning strategy presented below is based on regularized ERM formulations using non-negative smooth loss functions and strongly convex regularizers and is defined as follows.

Let $\ell : \mathcal{Y} \times \mathcal{Y} \to \mathbb{R}_+$ be an H-smooth loss function with $\mathcal{Y}$ an output space and let $\Omega : \mathcal{H} \to \mathbb{R}_+$ be a σ-strongly convex function with respect to a norm $\|\cdot\|$ and to a hypothesis space $\mathcal{H}$. Given a target training set $T = \{(\mathbf{x}_i, y_i)\}_{i=1}^m$, $\lambda \in \mathbb{R}_+$, n source hypotheses $\{h_{\text{src}}^i\}_{i=1}^n$ and a parameter vector $\boldsymbol{\beta}$ verifying $\Omega(\boldsymbol{\beta}) \leq \rho$, the transfer algorithm generates a target hypothesis $h_{\hat{\boldsymbol{w}},\boldsymbol{\beta}}$ so that

$$\hat{\boldsymbol{w}} = \underset{\mathbf{w} \in \mathcal{H}}{\text{argmin}} \left\{ \frac{1}{m} \sum_{i=1}^{m} \ell(\langle \boldsymbol{w}, \mathbf{x}_i \rangle + h_{\text{src}}^{\boldsymbol{\beta}}(\mathbf{x}_i), y_i) + \lambda \Omega(\mathbf{w}) \right\}.$$

In the formulation above, the loss function is only minimized with respect to $\mathbf{w}$ and not specifically with respect to $\boldsymbol{\beta}$. However, it is assumed that $\Omega(\boldsymbol{\beta}) \leq \rho$ making $\boldsymbol{\beta}$ constrained by a strongly convex function, which allows one to cover regularized algorithms that consider an additional regularization with respect to $\boldsymbol{\beta}$. As in the previous analysis, the key quantity $\mathrm{R}_{\mathcal{T}}(h_{\mathrm{src}}^{\boldsymbol{\beta}})$ measuring the relevance of the source hypothesis on the target domain will play a crucial role in the analysis of the generalization properties of $h_{\hat{\boldsymbol{w}},\boldsymbol{\beta}}$. To illustrate the types of algorithms covered by this analysis, we can consider the least squares based regularization. More formally, given a target training sample $T = \{(\mathbf{x}_i, y_i)\}_{i=1}^m$, source hypothesis $\{\mathbf{w}_{\mathrm{src}}^i\} \subset \mathcal{H}$, the parameters $\boldsymbol{\beta} \in \mathbb{R}^n$ and $\lambda \in \mathbb{R}_+$, the least squares algorithm with biased regularization outputs the target hypothesis

$$h(\mathbf{x}) := \langle \hat{\boldsymbol{w}}, \mathbf{x} \rangle,$$

where

$$\hat{\boldsymbol{w}} = \underset{\mathbf{w} \in \mathcal{H}}{\mathrm{argmin}} \left\{ \frac{1}{m} \sum_{i=1}^{m} (\langle \hat{\boldsymbol{w}}, \mathbf{x}_i \rangle - y_i)^2 + \lambda \| \mathbf{w} - \sum_{j=1}^{n} \beta_j \mathbf{w}_{\mathrm{src}}^j \|_2^2 \right\}. \qquad [7.2]$$

The problem defined by equation [7.2] can be interpreted as the minimization of the empirical error on the target sample, while keeping the solution close to the (best) linear combination of source hypotheses. It can actually be proved that this formulation is a special case of the classic regularized ERM [KUZ 17, KUZ 18]. While the formulation presented in equation [7.2] is limited to linear combination of source hypotheses living in the same space of the target predictor, it can be generalized allowing one to treat the source hypotheses as "black boxes" predictors [KUZ 17, KUZ 18].

The results presented below correspond to generalization bounds for regularized ERM-based algorithms.

THEOREM 7.3 ([KUZ 17]).– *Let $h_{\hat{\boldsymbol{w}},\boldsymbol{\beta}}$ a hypothesis output by a regularized ERM algorithm from a m-sized training set T i.i.d. from the target domain $\mathcal{T}$, n source hypotheses $\{h_{src}^i : \|h_{src}^i\|_\infty \leq 1\}_{i=1}^n$, any source weights $\boldsymbol{\beta}$ obeying $\Omega(\boldsymbol{\beta}) \leq \rho$ and $\lambda \in \mathbb{R}_+$. Assume that the loss is bounded by M: $\ell(h_{\hat{\boldsymbol{w}},\boldsymbol{\beta}}(\mathbf{x}), y) \leq M$ for any $(\mathbf{x}, y)$ and any training set. Then, denote $\kappa = \frac{H}{\sigma}$ and assuming that $\lambda \leq \kappa$ with probability at least $1 - e^{-\eta}$, $\forall \eta \geq 0$:*

$$\mathrm{R}_{\mathcal{T}}(h_{\hat{\boldsymbol{w}},\boldsymbol{\beta}}) \leq \mathrm{R}_{\hat{T}}(h_{\hat{\boldsymbol{w}},\boldsymbol{\beta}}) + \mathcal{O}\left(\frac{\mathrm{R}_{\mathcal{T}}^{src} \kappa}{\sqrt{m}\lambda} + \sqrt{\frac{\mathrm{R}_{\mathcal{T}}^{src} \rho \kappa^2}{m\lambda}} + \frac{M\eta}{m \log\left(1 + \sqrt{\frac{M\eta}{u^{src}}}\right)} \right)$$

$$\leq \mathrm{R}_{\hat{\mathcal{T}}}(h_{\hat{\mathbf{w}},\boldsymbol{\beta}}) + \mathcal{O}\left(\frac{\kappa}{\sqrt{m}}\left(\frac{\mathrm{R}_{\mathcal{T}}^{src}}{\lambda} + \sqrt{\frac{\mathrm{R}_{\mathcal{T}}^{src}\rho}{\lambda}}\right) + \frac{\kappa}{m}\left(\frac{\sqrt{\mathrm{R}_{\mathcal{T}}^{src}M\eta}}{\lambda} + \sqrt{\frac{\rho}{\lambda}}\right)\right),$$

where $u^{src} = \mathrm{R}_{\mathcal{T}}^{src}\left(m + \frac{\kappa\sqrt{m}}{\lambda}\right) + \kappa\sqrt{\frac{\mathrm{R}_{\mathcal{T}}^{src}m\rho}{\lambda}}$ *and* $\mathrm{R}_{\mathcal{T}}^{src} = \mathrm{R}_{\mathcal{T}}(h_{src}^{\boldsymbol{\beta}})$ *is the risk of the source hypothesis combination.*

The risk of the source hypothesis combination on the target domain (R^{src}) provides an important information that can be interpreted in two aspects: obviously the performance of the source hypothesis combination on the target domain, but also an indicator of the relatedness between the source and target domains:

– when the source information is bad for the target domain, i.e. $\mathrm{R}_{\mathcal{T}}^{\text{src}}$ is high, then $h_{\text{src}}^{\boldsymbol{\beta}}$ is useless for the transfer task. This can be interpreted as learning with no useful auxiliary information. Assuming that $\mathrm{R}_{\mathcal{T}}^{\text{src}} \leq M$, from theorem 7.3, we can get that $\mathrm{R}_{\mathcal{T}}(h_{\hat{\mathbf{w}},\boldsymbol{\beta}}) - \mathrm{R}_{\hat{\mathcal{T}}}(h_{\hat{\mathbf{w}},\boldsymbol{\beta}}) \leq \mathcal{O}(1/(\sqrt{m}\lambda))$. We recover classic rate suggesting that the approach is robust to negative transfer;

– when the source domain is informative for the target domain, one can have guarantees for small learning samples. In particular, let $m = \mathcal{O}(1/\mathrm{R}_{\mathcal{T}}^{\text{src}})$, then we can obtain a convergence rate of $\mathcal{O}(\sqrt{\rho}/m\sqrt{\lambda})$. This implies that a fast convergence rate behavior can be obtained with "small m" that depends on the performance of the combined source hypotheses. Asymptotically, the theorem also shows that a rate of $\mathcal{O}(\mathrm{R}_{\mathcal{T}}^{\text{src}}/(\sqrt{m}\lambda) + \sqrt{\mathrm{R}_{\mathcal{T}}^{\text{src}}\rho/m\lambda})$ can be obtained where the term $\mathrm{R}_{\mathcal{T}}^{\text{src}}$ is related to the constant factor of the rate. Hence, a small $\mathrm{R}_{\mathcal{T}}^{\text{src}}$ allows one to have a faster convergence making use of the information coming the source hypotheses combination;

– when the source domain is actually perfect for the target domain, *i.e.* $\mathrm{R}_{\mathcal{T}}^{\text{src}} = 0$, the source hypothesis is able to perfectly predict labels of instances of the target domain. In this case, theorem 7.3 implies that $\mathrm{R}(h_{\hat{\mathbf{w}},\boldsymbol{\beta}}) = \mathrm{R}_{\hat{\mathcal{T}}}(h_{\hat{\mathbf{w}},\boldsymbol{\beta}})$ with probability one. For most commonly used smooth loss functions, this setting is realistic only if source and target domains are the same and the task considered is noise free. Anyway, it is possible for some specific loss, such as the squared hinge loss, and with a target domain that can perfectly classified by the source.

7.3.4. *Comparison with standard theory of domain adaptation*

The seminal results from [BEN 10a] and [MAN 09a] have provided the first theoretical frameworks for domain adaptation using a domain divergence between distributions. Following the results presented in Chapter 3 on divergence-based generalization bounds derived in the literature, these bounds have in general the following form:

$$\mathrm{R}_{\mathcal{T}}(h) \leq \mathrm{R}_{\mathcal{S}}(h) + d(\mathcal{S}_{\mathbf{X}}, \mathcal{T}_{\mathbf{X}}) + \lambda,$$

where d is a divergence between source and target marginal distributions and λ refers to the adaptation capability of the hypothesis class $\mathcal{H}$ from where h is taken (e.g. the combined error of the ideal hypothesis h^* that minimizes $\mathrm{R}_{\mathcal{S}}(h) + \mathrm{R}_{\mathcal{T}}(h)$).

This type of bounds cannot be directly compared to result of theorem 7.3, even though the term R^{src} can be interpreted as the term corresponding to the $\mathcal{H}\Delta\mathcal{H}$-divergence. Indeed, we note that by defining $\mathcal{H} = \{\mathbf{x} \mapsto \langle \boldsymbol{\beta}, \boldsymbol{h}_{\text{src}}(\mathbf{x}) \rangle | \ \Omega(\boldsymbol{\beta}) \leq \tau\}$ where $\boldsymbol{h}_{\text{src}}(\mathbf{x}) = [h^1_{\text{src}}(\mathbf{x}), \ldots, h^n_{\text{src}}(\mathbf{x})]^\top$, and fixing $h = h^{\boldsymbol{\beta}}_{\text{src}} \in \mathcal{H}$, we can write

$$R^{\text{src}} = \mathrm{R}_{\mathcal{T}}(h^{\boldsymbol{\beta}}_{\text{src}}) \leq \mathrm{R}_{\mathcal{S}}(h^{\boldsymbol{\beta}}_{\text{src}}) + d_{\mathcal{H}\Delta\mathcal{H}}(\mathcal{S}_{\mathbf{X}}, \mathcal{T}_{\mathbf{X}}) + \lambda_{\mathcal{H}}.$$

If we plug this inequality into the result presented above, we have for any hypothesis h and $\lambda \leq 1$, $\rho \leq 1/\lambda$ the following:

$$\mathrm{R}_{\mathcal{T}}(h) \leq \mathrm{R}_{\hat{\mathcal{S}}}(h) + \mathcal{O}\left(\frac{\mathrm{R}_{\mathcal{S}}(h^{\boldsymbol{\beta}}_{\text{src}}) + d_{\mathcal{H}\Delta\mathcal{H}}(\mathcal{S}_{\mathbf{X}}, \mathcal{T}_{\mathbf{X}}) + \lambda_{\mathcal{H}}}{\sqrt{m}\lambda} + \frac{1}{m\lambda}\right). \qquad [7.3]$$

The two results agree on the fact that the divergence between the domains has to be small to generalize well. The divergence is actually controlled by the choice of $\boldsymbol{h}_{\text{src}}$ and by the complexity of the hypothesis class $\mathcal{H}$ controlled by τ. In classic domain adaptation, a hypothesis h performs well on the target domain only if it performs well on the source domain under the condition that $\mathcal{H}$ is expressive enough for ensuring adaptation or, in other words, that the λ term should be small. In HTL, however, this condition can be relaxed as highlighted by equation [7.3] implying that a good source model has to perform well on its own domain. Additionally, while in classic domain adaptation the λ-term is assumed to be small – otherwise there is no hypothesis able to perform well on both domains at the same time and adaptation cannot be effective – in HTL transfer can still be effective even for large λ but this is due to the supervised aspect of the setting used [KUZ 17].

Note that HTL has been also investigated from Bayesian perspective through the approach of Li and Bilmes [LI 07] where the authors propose PAC–Bayesian study of HTL and derive bounds capturing the relationship between domains by an additive KL-divergence term, which is classic in the PAC-Bayesian setting. In the particular case of logistic regression, they showed that the divergence term is upper bounded by $\|h - h_{\text{src}}\|^2$, motivating the biased regularization term in logistic regression and the interest of incorporating source hypothesis to the adaptation model. We invite the interested reader to have a look to the related publications.

7.3.5. *HTL for metric learning*

The notion of metric – used here as a generic term for distance, similarity or dissimilarity – allows one to compare data that can be represented in various heterogeneous formats such as numerical vectors, structured objects like strings, trees and graphs, or temporal series. Based on the saying "Birds of a feather flock together", metrics play a crucial role in a large variety of supervised or unsupervised learning methods, such as the widely used k-nearest neighbors (k-NN), kernel-based methods in classification, the k-means algorithm in clustering and in other domains such as information retrieval, ranking algorithms or recommendation systems to name a few. The use of an appropriate metric is key for all these methods. A classic idea is to focus on metrics that can be tailored to a problem at hand, or in other words, metrics that are parameterizable by nature (e.g. Mahalanobis distance, edit distance and cosine similarity). Since manually tuning metrics for a given real-world problem is often difficult and tedious, a lot of work has gone into automatically learning them from training data, leading to the emergence of *metric learning* for which the reader can find many interesting surveys [BEL 13, KUL 13, BEL 15]. So far, classical settings are based on supervised knowledge mainly expressed in the form of constraints over labels. Generally speaking, supervised metric learning approaches rely on the reasonable intuition that a good similarity (or distance) function should assign a large (respectively small) score to pairs of points of the same class (respectively different class), as illustrated in Figure 7.1. Following this idea, they aim at finding the parameters (usually a matrix) of a given metric so that it best satisfies local (pair or triplet)-based constraints built from the training sample. This usually boils down to geometrically moving closer examples with the same label and putting far away instances of different classes. Therefore, metric learning gathers in some way notions of similarity, geometry and discriminative power.

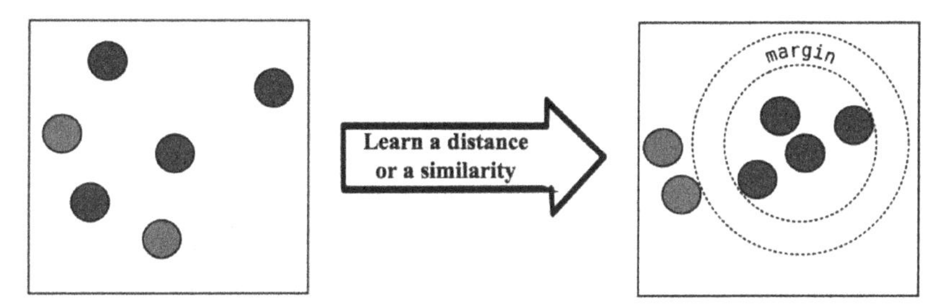

Figure 7.1. *Intuition behind metric learning. Before learning (left pane), red and blue points are not well separated. After learning (right pane), red and blue points are separated by a certain margin. For a color version of this figure, see www.iste.co.uk/redko/domain.zip*

During the past decade, Mahalanobis distance learning has attracted a lot of interest. The Mahalanobis distance between two d-dimensional numerical vectors $\mathbf{x}$ and $\mathbf{x}'$ is defined by $d^2(\mathbf{x}, \mathbf{x}') = (\mathbf{x} - \mathbf{x}')^T \mathbf{M}(\mathbf{x} - \mathbf{x}')$, where $\mathbf{M}$ is a $d \times d$

dimension matrix[1]. The interest for learning such a distance (i.e. the matrix $\mathbf{M}$) is two-fold: first, using Cholesky decomposition, one can rewrite $\mathbf{M}$ as $\mathbf{M} = \mathbf{L}^T\mathbf{L}$, where $\mathbf{L} \in \mathbb{R}^{k \times d}$ and k is the rank of $\mathbf{M}$. This means that learning a Mahalanobis distance implicitly corresponds to computing the Euclidean distance after a (learned) linear projection of the data defined by L. Indeed, $d^2(\mathbf{x}, \mathbf{x}')$ can be rewritten as $d^2(\mathbf{x}, \mathbf{x}') = (\mathbf{L}\mathbf{x} - \mathbf{L}\mathbf{x}')^T(L\mathbf{x} - L\mathbf{x}')$. By this way, we aim at improving via a suitable linear projection the discriminative power of the Euclidean metric. This approach can also be extended to learn linear or nonlinear manifolds. Second, by incorporating knowledge from training data (in the form of pairs or triplets-based constraints) in matrix $\mathbf{M}$, one can learn ad hoc metrics, which are relevant for the task at hand and performing well on unseen data when plugged, for example, into nearest neighbors-based algorithms.

Using the same idea as presented in the previous section, some metric learning approaches make use of auxiliary knowledge [BOH 14, CAO 13, DAV 07, PAR 10, PER 15a, ZHA 09] in order to help to learn $\mathbf{M}$ efficiently, especially when only *few labeled data* are available. This is, in particular, the case for *supervised regularized metric learning approaches* where the regularizer is *biased* with respect to an auxiliary metric given under the form of a matrix. The main objective here is to make use of this *a priori* knowledge in a setting where only few labeled data are available to help learning. For example, in the context of learning a positive semidefinite (PSD) matrix $\mathbf{M}$ plugged into a Mahalanobis-like distance as discussed above, let $\mathbf{I}$ be the identity matrix used as an auxiliary knowledge and$\|\mathbf{M} - \mathbf{I}\|$ is a biased regularizer often considered. This regularization can be interpreted as follows: learn $\mathbf{M}$ while trying to stay close to the Euclidean distance, or from another standpoint try to learn a matrix $\mathbf{M}$ that performs better than $\mathbf{I}$. Other standard matrices can be used such as the inverse of the variance–covariance matrix Σ^{-1}. Note that if we take the $\mathbf{0}$ matrix, we retrieve the classical unbiased regularization term.

Another useful setting comes when $\mathbf{I}$ is replaced by any auxiliary matrix $\mathbf{M}_{\mathcal{S}}$ learned from another task. This corresponds to a *transfer learning* approach where the biased regularization can be interpreted as transferring the knowledge brought by $\mathbf{M}_{\mathcal{S}}$ for learning $\mathbf{M}$. This setting is called *Metric Hypothesis Transfer Learning* [PER 15b] and is a particular case of the HTL setting seen in the previous section specialized to *metric learning*.

In this section, we present theoretical results tailored to the problem of learning a new metric with the help of biased regularization involving an auxiliary metric. First, we introduce several notations and definitions that are used throughout the section. Let

1 Note that this distance is a generalization of some well-known distances: when $\mathbf{M} = \mathbf{I}$, $\mathbf{I}$ being the identity matrix, we retrieve the Euclidean distance, when $\mathbf{M} = \Sigma^{-1}$ where Σ is the variance–covariance matrix of the data at hand, it actually corresponds to the original definition of a Mahalanobis distance.

us consider metrics corresponding to distance functions $\mathbf{X} \times \mathbf{X} \to \mathbb{R}^+$ parameterized by a $d \times d$ PSD matrix $\mathbf{M}$ denoted by $\mathbf{M} \succeq 0$. In the following, a metric is represented by its matrix $\mathbf{M}$. We also consider that we have access to some additional information under the form of an auxiliary $d \times d$ matrix $\mathbf{M}_{\mathcal{S}}$ that we call source metric or source hypothesis in the following. We denote the Frobenius norm by $\|\cdot\|_{\mathcal{F}}$; $\mathbf{M}_{kl}$ represents the value of the entry at index (k, l) in matrix $\mathbf{M}$, $[a]_+ = \max(a, 0)$ denotes the hinge loss and $[n]$ the set $\{1, \ldots, n\}$ for any $n \in \mathbb{N}$.

Let $T = \{\mathbf{z}_i = (\mathbf{x}_i, y_i)\}_{i=1}^n$ be a labeled training set drawn from the target domain $\mathcal{T}$. We consider the following optimization problem for *biased regularized metric learning*

$$\mathbf{M}^* = \underset{\mathbf{M} \succeq 0}{\operatorname{argmin}}\ L_T(\mathbf{M}) + \lambda \|\mathbf{M} - \mathbf{M}_{\mathcal{S}}\|_{\mathcal{F}}, \qquad [7.4]$$

where $L_T(\mathbf{M}) = \frac{1}{n^2} \sum_{\mathbf{z},\mathbf{z}' \in T} \ell(\mathbf{M}, \mathbf{z}, \mathbf{z}')$ stands for the empirical risk of a metric hypothesis $\mathbf{M}$. Similarly, we denote the true risk by $L_{\mathcal{T}}(\mathbf{M}) = \underset{\mathbf{z},\mathbf{z}' \sim \mathcal{T}}{\mathbf{E}} \ell(\mathbf{M}, \mathbf{z}, \mathbf{z}')$. In this part, we only consider convex, k-lipschitz and (σ, m)-admissible losses for which we recall the definitions below.

DEFINITION 7.3 (k-lipschitz continuity).– *A loss function $\ell(\mathbf{M}, \mathbf{z}, \mathbf{z}')$ is k-lipschitz w.r.t. its first argument if, for any matrices $\mathbf{M}$, $\mathbf{M}'$ and any pair of examples $\mathbf{z}, \mathbf{z}'$, there exists $k \geq 0$ such that*

$$|\ell(\mathbf{M}, \mathbf{z}, \mathbf{z}') - \ell(\mathbf{M}', \mathbf{z}, \mathbf{z}')| \leq k \|\mathbf{M} - \mathbf{M}'\|_{\mathcal{F}}.$$

This property ensures that the loss deviation does not exceed the deviation between matrices $\mathbf{M}$ and $\mathbf{M}'$ with respect to a positive constant k.

DEFINITION 7.4 ((σ, m)-admissibility).– *A loss function $\ell(\mathbf{M}, \mathbf{z}, \mathbf{z}')$ is (σ, m)-admissible, w.r.t. $\mathbf{M}$, if it is convex w.r.t. its first argument and if for any two pairs of examples $\mathbf{z}_1, \mathbf{z}_2$ and $\mathbf{z}_3, \mathbf{z}_4$, we have*

$$|\ell(\mathbf{M}, \mathbf{z}_1, \mathbf{z}_2) - \ell(\mathbf{M}, \mathbf{z}_3, \mathbf{z}_4)| \leq \sigma \, |y_1 y_2 - y_3 y_4| + m,$$

where $y_i y_j = 1$ if $y_i = y_j$ and -1 otherwise. Thus, $|y_1 y_2 - y_3 y_4| \in \{0, 2\}$.

This property bounds the difference between the losses of two pairs of examples by a value only related to the labels, plus a constant independent from $\mathbf{M}$.

The theoretical results derived from [PER 15b] make use of an extension of the notion of *algorithmic stability* (see subsection 1.2.4) to metric learning, which allows one to provide generalization guarantees for a learned metric. As we have explained in the previous chapters, a learning algorithm is stable if a slight modification in its input does not change its output much. Similar to the original work, the authors define a notion of *on-average-replace-two-stability* that corresponds to an adaptation of the notion of *on-average-replace-one-stability* to metric learning proposed in [SHA 14]. Its definition is given below.

DEFINITION 7.5 (On-average-replace-two-stability).– *Let* $\epsilon : \mathbb{N} \rightarrow \mathbb{R}$ *be monotonically decreasing and let* $U(n)$ *be the uniform distribution over* $[n]$. *A metric learning algorithm is on-average-replace-two-stable with rate* $\epsilon(n)$ *if for every distribution* $\mathcal{T}$

$$\mathop{\mathbf{E}}_{\substack{T\sim\mathcal{T}^n\\ i,j\sim U(n)\\ \mathbf{z}_1,\mathbf{z}_2\sim\mathcal{T}}} \left[\ell(\mathbf{M}^{ij^*}, \mathbf{z}^i, \mathbf{z}^j) - \ell(\mathbf{M}^*, \mathbf{z}^i, \mathbf{z}^j)\right] \leq \epsilon(n)$$

where $\mathbf{M}^*$, *respectively* $\mathbf{M}^{ij^*}$, *is the optimal solution when learning with the training set* T, *respectively* T^{ij}. T^{ij} *is obtained by replacing* $\mathbf{z}^i$, *the* ith *example of* T, *by* $\mathbf{z}_1$ *to get a training set* T^i *and then by replacing* $\mathbf{z}^j$, *the* jth *example of* T^i, *by* $\mathbf{z}_2$.

This property ensures that, given two examples involved in a pair, learning with or without them do not imply a big change in the metric prediction. Note that the property is required to be true on average over all the possible training sets of size n. When this definition is true, it implies $\mathop{\mathbf{E}}_{T\sim\mathcal{T}} \left[L_{\mathcal{T}}(\mathbf{M}^*) - L_T(\mathbf{M}^*)\right] \leq \epsilon(n)$. The next theorem shows that a metric learning algorithm based on Problem [7.4] is on-average-replace-two-stable.

THEOREM 7.4 (On-average-replace-two-stability [PER 15b]).– *Given a training sample* T *of size* n *drawn i.i.d. from* $\mathcal{T}$, *a metric learning algorithm based Problem* [7.4] *is on-average-replace-two-stable with* $\epsilon(n) = \frac{8k^2}{\lambda n}$.

We can now bound the expected true risk of biased metric learning algorithms based on problem [7.4].

THEOREM 7.5 (On average bound [PER 15b]).– *For any convex, k-lipschitz loss, we have:*

$$\mathop{\mathbf{E}}_{T\sim\mathcal{T}^n} \left[L_{\mathcal{T}}(\mathbf{M}^*)\right] \leq L_{\mathcal{T}}(\mathbf{M}_{\mathcal{S}}) + \frac{8k^2}{\lambda n}$$

where the expected value is taken over size-n *training sets.*

This bound shows that with a sufficient number of examples with respect to a fast convergence rate given by $\mathcal{O}(1/n)$, we obtain on average a metric that is at least as good as the source hypothesis. Thus, choosing a good source metric is vital to learn well in general and in case when only a small learning sample is available.

The second result is based on an adaptation of the classic framework of *uniform stability* for metric learning as proposed in [JIN 09, BEL 15] where only example is changed in the learning sample from which the learning pairs are built. Assuming that supremum over the worst-case possible pair of the difference between the losses of two possible learned metrics behaves as $\mathcal{O}\left(\frac{1}{n}\right)$, one can prove a generalization guarantee based on the general algorithmic stability setting [BOU 02] recalled in subsection 1.2.4. In this context, it is possible to derive a result of the following form:

$$L_{\mathcal{T}}(\mathbf{M}) \leq L_{T}(\mathbf{M}) + \mathcal{O}\left(\frac{1}{\sqrt{n}}\right)\Big(f(L_{\mathcal{T}}(\mathbf{M}_{\mathcal{S}}), \|\mathbf{M}_{\mathcal{S}}\|_{\mathcal{F}}, \lambda, c_{\ell})\Big), \qquad [7.5]$$

where f is decreasing with $L_{\mathcal{T}}(\mathbf{M}_{\mathcal{S}})$, and c_{ℓ} represents a constant related to the loss function used. The second term of the left-hand side of equation [7.5] means that as the number of examples available for learning increases, the quality of the source metric is of decreasing importance. As mentioned previously, this result is similar to other results in domain adaptation or HTL [BEN 10a, KUZ 13] where it was shown that with the increasing number of target examples, the necessity of having source examples to learn a low error hypothesis decreases.

7.4. Summary

In this chapter, we presented theoretical results that allow to take into consideration algorithmic properties of adaptation algorithms. First, we recalled how the algorithmic robustness can be extended to the domain adaptation setting where one assumes a relaxation of the covariate-shift assumption. Second, we focused on a different domain adaptation setting called HTL, where one does not have access to source samples, but to source model(s) given by the learned hypotheses. In this setting, we presented theoretical results obtained in the case of regularized ERM-based algorithms. Finally, this chapter concludes with a brief presentation of theoretical results based on the algorithmic stability for a large family of metric learning algorithms in the HTL setting.

In general, we may highlight several important differences of this framework with respect to the results seen in the previous chapters. They are the following:

1) Contrary to the divergence-based bounds, the learning guarantees presented in this chapter do not include a term measuring the discrepancy between the marginal distributions of the two domains. This is rather expected, as in HTL scenario we do not have access to a learning sample from the source domain, but only to a hypothesis learned on it.

2) The potential success of adaptation in the HTL framework depends on the performance of the source hypothesis on the target distribution and allows to learn a better hypothesis, even on small samples when some assumptions are fulfilled.

3) The theoretical results presented for metric learning algorithms show that even when labeled target data are available, having an *a priori* knowledge given by the source model can improve the overall performance. This result is somewhat similar to the bound obtained for the combined error in Chapter 3, where it was proved that a mixture of source and target data can sometimes improve the performance compared to learning with target data only.

Finally, we also note that this chapter presents theoretical results for particular learning settings such as, for instance, metric learning-based adaptation algorithms. This is contrary to general divergence-based results that hold true, for no matter what algorithm is used. In the following chapter, we give another example of adaptation algorithm is for which one can obtain substantially different theoretical results with independent insights. The considered algorithms are self-iterative methods that present a subject of investigation with many interesting theoretical properties.

8

Iterative Domain Adaptation Methods

Self-iterative strategies consist of labeling (unlabeled) target data using a hypothesis built from the available source examples in a step-by-step fashion such that at each step the classifier is learned on a modified sample that includes the pseudo-labeled (or semilabeled) examples from the previous step. In this scenario, a new classifier is trained in way that allows us to take into account the information provided by both domains with a stopping point corresponding to a state where all target instances are labeled. This iterative process, taking its origins from self-training or cotraining strategies [BLU 98], has led to several strategies for unsupervised domain adaptation with the most notorious examples being given by the DASVM algorithm [BRU 10] and the approach of Perez and Sánchez-Montañés [PÉR 07] based on the expectation–maximization algorithm. More recently, several approaches have also extended these ideas based on different algorithmic principles with regularization information coming from pseudo-labeled data (e.g. [LEE 13, SEN 16, SAI 17]).

Despite a very intuitive nature, these unsupervised domain adaptation methods are known, however, to have several challenging bottlenecks related to the following problems:

– they rely on defining the target instances that should be pseudo-labeled at each step and the type of pseudo-labels used;

– one has to choose the source labeled instances that need to be removed from the learning sample when appropriate;

– at each step, one has to define the number of pseudo-labeled instances that will be added to the learning sample used in the next iteration.

While there exist different algorithmic strategies with established theoretical guarantees, there are almost no theoretical frameworks for domain adaptation in self-iterative scenarios. In this chapter, we propose to present some of the few existing results, accordingly structured with respect to two main directions: iterative self-labeling based methods and a short description of boosting-based results.

8.1. Iterative self-labeling domain adaptation

The work presented in [HAB 13b] is one of the few published theoretical studies about the iterative self-labeling procedure for domain adaptation. This study focuses on a simple self-labeling method that (given a hypothesis class, a learning algorithm for a particular iteration, a labeled source sample S and an unlabeled set of target instances T_u) proceeds by performing two procedures at each step:

1) it learns a model from the current sample S;

2) it pseudo-labels some target samples from T_u by the model from the previous step and incorporates them into the source sample S to progressively modify the current classifier.

The procedure is repeated until all the target instances have been processed or when a stopping criterion has been reached. Sometimes, an additional procedure aiming at removing some original source instances from S is performed. An application of this iterative strategy can be found in the DASVM algorithm [BRU 10], the objective of which, is to progressively adapt the learned classifier to the data coming from the target domain.

Since the true labels of the target examples are not available, the choice of the target instances to be (pseudo-) labeled and the design of the domain adaptation algorithm need special care in order to prevent negative transfer phenomena resulting from possible successive mislabelings. In this section, we present the necessary conditions required for such iterative domain adaptation algorithms to solve an actual domain adaptation problem. This theoretical study is done in the limited case where a random selection of the pseudo-labeled target examples is performed at each iteration. While this setup is certainly not optimal, it is sufficient to define a simple framework for which we can provide minimal guarantees that an efficient iterative domain adaptation algorithm should fulfill. This framework is inspired by the famous ADABOOST algorithm [FRE 97] and relies on a notion of weak domain adaptation assumption. To this end, we first begin by recalling the definition of *weak learner* presented in [FRE 96].

DEFINITION 8.1 (Weak learner [FRE 96]).– *A classifier $h^{(i)}$ learned at time i from the current training set $S^{(i)}$ is a weak learner if its empirical error $\mathrm{R}_{S^{(i)}}(h^{(i)})$ over $S^{(i)}$ is smaller than $\frac{1}{2}$, i.e.*

$$\mathrm{R}_{S^{(i)}}(h^{(i)}) = \underset{\mathbf{x}_l^S \in S^{(i)}}{\mathbf{E}} [h^{(i)}(\mathbf{x}_l^S) \neq y_l^S]$$
$$= \frac{1}{2} - \gamma_S^{(i)},$$

where $\gamma_S^{(i)} > 0$ measures to what extent the predictions obtained with $h^{(i)}$ are better than random guessing.

In order to extend this definition to iterative domain adaptation, the notion of a *weak domain adaptation learner* with respect to pseudo-labeled target data needs to be proposed. As this definition requires the notion of pseudo-labeling, we give it as follows.

DEFINITION 8.2 ([HAB 13b]).– *A pseudo-labeled target point inserted in $S^{(i)}$ at step i in place of a source point is an example randomly drawn from T_u (without replacement) labeled by the hypothesis $h^{(i-1)}$ learned from the previous training set $S^{(i-1)}$.*

The notion of *weak domain adaptation learner* evoked before is presented in the following definition.

DEFINITION 8.3 (Weak domain adaptation learner [HAB 13b]).– *A classifier $h^{(i)}$ learned at step i from the current training set $L_S^{(i)}$ is a weak domain adaptation learner with respect to a set $SL_j = \{\mathbf{x}_1^T, \ldots, \mathbf{x}_{2k}^T\}$ of $2k$ pseudo-labeled target data inserted at step j if its empirical error $\mathrm{R}_{SL_j}(h^{(i)})$ over these $2k$ points – according to their true (unknown) label – is smaller than $\frac{1}{2}$, i.e.*

$$\mathrm{R}_{SL_j}(h^{(i)}) = \underset{\mathbf{x}_l^T \in SL_j}{\mathbf{E}} [h^{(i)}(\mathbf{x}_l^T) \neq y_l^T] < \frac{1}{2}.$$

From this definition, a first necessary condition on the error of a weak domain adaptation learner can be derived.

THEOREM 8.1 ([HAB 13b]).– *Let $h^{(i)}$ be a weak hypothesis learned at step i from $S^{(i)}$, and $\tilde{\mathrm{R}}_S^{(i)}(h^{(i)}) = \frac{1}{2} - \gamma_S^{(i)}$ its corresponding empirical error*[1]*. Let $\mathrm{R}_{\hat{T}}(h^{(i)}) = \frac{1}{2} - \gamma_T^{(i)}$ be the (unknown) empirical error of $h^{(i)}$ over the target sample T_u. $h^{(i)}$ is a weak domain adaptation learner w.r.t. a set $SL_j = \{\mathbf{x}_1^T, \ldots, \mathbf{x}_{2k}^T\}$ of $2k$ pseudo-labeled target data inserted at step j ($j \leq i$) if $\gamma_T^{(j-1)} > 0$.*

1 In this self-labeling domain adaptation context, the tilde symbol (˜) is used instead of the usual notations of empirical risk (i.e. $\mathrm{R}_{\hat{S}}$ and R_S) as for the examples of $S^{(i)}$ coming from T_u, we only have access to their pseudo-label which can be wrong as expressed in the proof. We also put the superscript of the sample $S^{(i)}$ as a superscript of the risk to avoid overcharged notations.

The proof of theorem 8.1 is given in Appendix 5. One may note that this theorem is simple to interpret. Indeed, it means that $h^{(i)}$ will be able to correctly classify (with respect to their unknown true label) more than k pseudo-labeled target examples among $2k$ if at least half of them have been correctly pseudo-labeled.

One may note that in the context of iterative algorithms considered in this chapter, solving a domain adaptation task consists of producing a hypothesis $h^{(\frac{N_S}{2k})}$ from pseudo-labeled target examples that outperforms a hypothesis $h^{(0)}$ learned only from source labeled data, i.e. $\mathrm{R}_{S}^{(\frac{N_S}{2k})}(h^{(\frac{N_S}{2k})}) < \mathrm{R}_{\hat{\mathcal{T}}}(h^{(0)})$. We will now give the necessary conditions required for an iterative domain adaptation algorithm that infers weak domain adaptation hypotheses to solve a domain adaptation task in the sense defined above.

THEOREM 8.2 ([HAB 13b]).– *Let $S^{(0)}$ be the original learning set made of N_S labeled source data and T_u a set of $N_T \geq N_S$ unlabeled target examples. Let $\mathcal{A}$ be an iterative domain adaptation algorithm that randomly changes, at each step i, $2k$ original source labeled points from $S^{(i)}$ by $2k$ pseudo-labeled target examples randomly drawn from T_u without replacement and infers at each step a weak domain adaptation hypothesis (according to definition 8.3). Let $h^{(\frac{N_S}{2k})}$ be the weak domain adaptation hypothesis learned by $\mathcal{A}$ after $\frac{N_S}{2k}$ such iterations needed to change $S^{(0)}$ into a new learning set made of only target examples. Algorithm $\mathcal{A}$ solves a domain adaptation task with $h^{(\frac{N_S}{2k})}$ if*

$$\gamma_S^{(i)} \geq \gamma_T^{(i)}, \quad \forall i = 1, \dots, \frac{N_S}{2k}, \qquad [8.1]$$

$$\gamma_S^{max} > \sqrt{\frac{\gamma_T^{(0)}}{2}}, \text{ where } \gamma_S^{max} = \max\left(\gamma_S^{(0)}, \dots, \gamma_S^{(n)}\right). \qquad [8.2]$$

The proof of theorem 8.2 is given in Appendix 5. Before analyzing this theorem, let us interpret via theorem 8.3 the meaning of condition [8.1].

THEOREM 8.3 ([HAB 13b]).– *If condition [8.1] of theorem 8.2 is verified, then for all i,*

$$\tilde{\mathrm{R}}_S^{(i+1)}(h^{(i)}) > \tilde{\mathrm{R}}_S^{(i)}(h^{(i)}).$$

The proof of theorem [8.3] is given in Appendix 5. As in boosting, one can interpret theorem 8.3 as the requirement for a classifier built from $S^{(i+1)}$ to learn something new about the target distribution $\mathcal{T}$. Note that in the classic boosting algorithm ADABOOST [FRE 96], the sufficient condition is rather $\mathrm{R}_S^{(i+1)}(h^{(i)}) = \frac{1}{2}$, as ADABOOST performs a linear combination of all the learned weak classifiers.

Since iterative domain adaptation algorithms only keep the last induced hypothesis, the condition $\mathrm{R}_S^{(i+1)}(h^{(i)}) = \frac{1}{2}$ is thus not sufficient. To sum up, by constraining the old classifier $h^{(i)}$ to work well on the new training set $L_S^{(i+1)}$, theorem 8.3 means that the domain adaptation algorithm has to infer a new hypothesis that is able to learn something new about $\mathcal{T}$. Note, however, that condition [8.1] alone is not sufficient to perform an actual adaptation. In theorem 8.2, condition [8.2] expresses the idea that, in addition to condition [8.1], at least one hypothesis has to significantly outperform the behavior of the classifier $h^{(0)}$ (learned only from source data) over L_T. Figure 8.1 illustrates these two conditions.

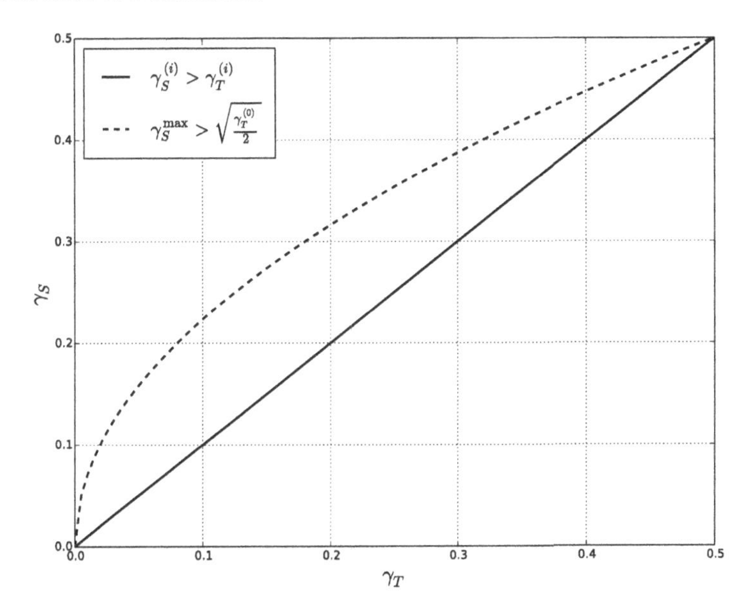

Figure 8.1. *Conditions for a domain adaptation algorithm to adapt well. The dashed line means that, according to condition [8.1] of theorem 8.2, γ_S must be greater than γ_T to learn something new at each iteration. The solid line expresses the additional condition [8.2] implying that γ_S^{max} must also be greater than $\sqrt{\frac{\gamma_T^{(0)}}{2}}$ (i.e. the area above the curve)*

The underlying intuitive information provided by the previous theoretical results is that an iterative domain adaptation algorithm $\mathcal{A}$ adapts well if at each step i the induced classifier $h^{(i)}$ satisfies the following conditions:

– the classifier $h^{(i)}$ must work well on T_u: the hypothesis $h^{(i)}$ learned by $\mathcal{A}$ from the source domain has to perform reasonably well in the target domain. This intuitive idea is formally expressed by the weak domain adaptation assumption of definition 8.3 and theorem 8.1 ($\gamma_T^{(i)} > 0$);

– the classifier $h^{(i)}$ must work well on S: as the quality of the target data inserted in the learning set mainly depends on the pseudo-labels assigned by $h^{(i)}$, one has to be confident in $h^{(i)}$. Therefore, $h^{(i)}$ has to perform sufficiently well on the data it has been learned from. This idea is formally expressed by condition [8.1] of theorem 8.2 ($\gamma_S^{(i)} > \gamma_T^{(i)}$);

– the algorithm $\mathcal{A}$ works better than a non-adaptation process: indeed, the final classifier $h^{(\frac{N_S}{2k})}$ learned only from pseudo-labeled target data has to perform better on T_u than a classifier learned only from source data. This idea is described by condition [8.2] of theorem 8.2 $\left(\gamma_S^{\max} > \sqrt{\frac{\gamma_T^{(0)}}{2}}\right)$ and illustrated by Figure 8.1.

A second remark about this theoretical study is that in the proof of theorem 8.2, the authors use $\gamma_S^{\max}$ to bound every $\gamma_S^{(j)}$ in order to make a transition from equations [A5.1] to [A5.2]. It is worth noting that one can use another strategy that can lead to a slightly different result. Coming back to equation [A5.1], we get:

$$
\begin{aligned}
\frac{4k}{N_S}\gamma_S^{(\frac{N_S}{2k})}\sum_{j=1}^{\frac{N_S}{2k}}\gamma_S^{(j-1)} &> \gamma_T^{(0)}, \text{ because of condition [8.1].}\\
\Leftrightarrow 2\gamma_S^{(\frac{N_S}{2k})}\frac{2k}{N_S}\sum_{j=1}^{\frac{N_S}{2k}}\gamma_S^{(j-1)} &> \gamma_T^{(0)}\\
\Leftrightarrow 2\gamma_S^{(\frac{N_S}{2k})}\bar{\gamma}_S &> \gamma_T^{(0)} \text{ where } \bar{\gamma}_S = \tfrac{2k}{N_S}\textstyle\sum_{j=1}^{\frac{N_S}{2k}}\gamma_S^{(j-1)}\\
\Leftrightarrow \bar{\gamma}_S &> \gamma_T^{(0)}, \text{ because } \forall j, \gamma_S^{(j)} < \tfrac{1}{2}.
\end{aligned} \qquad [8.3]
$$

Equation [8.3] provides a different condition from that given in condition [8.2]. By using the average of the γ values rather than the maximum value, equation [8.3] provides a less constraining condition that has to be fulfilled to ensure efficient adaptation. Indeed, while an upper bound on $\frac{N_S}{2k}$ terms was used to obtain equation [A5.2] from equation [A5.1], equation [8.3] was obtained with only one upper bound.

Finally, one may note that throughout this analysis, the quantity γ_T is used to provide theoretical guarantees of an iterative domain adaptation algorithm. It should be noted that in practice $\gamma_T^{(i)}$ is unknown because T_u is composed of unlabeled examples only. However, these results provide important information about the strategy that can be used to adapt well and to efficiently select the target points. For example, DASVM selects k examples falling into the margin band that are the closest to the margin bounds on each side of the linear separator. In the light of this theoretical study, this strategy tends to satisfy the required conditions to adapt well. Indeed, such selected points are those with the highest probability of modifying the model produced by the SVM algorithm at the next step, as they are likely to be associated with non-zero Lagrange multipliers and thus allow to learn something new about the target domain. On the other hand, by choosing those points that are the closest to the margin bounds, i.e. the farthest from the hyperplane, one increases the probability of the pseudo-labeled data being correctly classified. Therefore, DASVM also tends to satisfy the weak domain adaptation assumption.

8.2. Boosting-based frameworks

In the previous section, the notion of a weak learner borrowed from boosting was adapted to obtain theoretical results for iterative self-labeling methods. Originally, the classic boosting approach is an ensemble method that, given a labeled learning sample, aims at inferring a set of (weak) classifiers $h_1, \ldots, h_N$, the individual decisions of which are combined to classify a new example. These classifiers are learned iteratively by means of an update rule that increases (respectively, decreases) the weights of instances that have been misclassified (respectively, correctly classified) by previous classifiers. In general, a weighted linear combination is performed to define the final decision function from the learned classifiers given as follows:

$$F^N(x) = \sum_{i=1}^{N} \alpha_i h_i(x),$$

where each α_i is a learned weight that represents the importance of the associated classifier in the decision function. In the domain adaptation setting, the adaptation of boosting is not straightforward as one has to deal with two domains, each are associated with its own data sample. Nevertheless, several domain adaptation approaches based on boosting have been proposed in the literature [DAI 07, HAB 13a, BEC 13, HAB 16]. In this section, we focus on two particular approaches that are known to benefit from established theoretical guarantees, namely:

– TRADABOOST [DAI 07], a method based on the classic boosting approach that requires labeled samples from both the source and target domains;

– SLDAB [HAB 16], a method that stands in the unsupervised domain adaptation framework where the algorithm takes as input a labeled source sample and an unlabeled target sample and makes use of the notion of a weak learner that specialized to a domain adaptation setting.

We start by presenting the TRADABOOST approach in section 8.2.1.

8.2.1. TRADABOOST *with both labeled source and target data*

In the domain adaptation context, the approach implemented by the algorithm TRADABOOST [DAI 07] takes as input a target labeled learning sample T and a source sample S, also called the auxiliary sample. Here, the target sample is supposed to be insufficient to train a good classifier, so the idea is to try to transfer some useful information from S to help learning in T. The proposed approach consists of reweighting the target examples at each iteration following a strategy similar to that of traditional boosting, with an additional step related to the adaptation of source labeled instances. More precisely, TRADABOOST proceeds as follows:

1) increase the weights of misclassified instances in T as in traditional boosting;

2) decrease the weights of misclassified data in S by multiplying the previous weight by a quantity $\beta^{|h_t(\mathbf{x}_i)-y_i|}$, where $\beta \in [0,1]$ and $(\mathbf{x}_i, y_i)$ is a labeled source example. The idea here is to try to keep only those source examples that are close and well aligned with the target sample.

Using this strategy, in the next round the misclassified examples from the source sample will have a smaller impact on the learning of a classifier. The intuition behind this process is to assume that these misclassified source examples are distributed differently enough from the target distribution and thus we should discard them. At the end of this procedure, source examples that are the most shifted with respect to the target distribution tend to have small weights and do not significantly affect the learned hypothesis, while those similar to the target domain gain more importance as they are expected to help with learning efficiently on target data. As the algorithm makes use of ADABOOST, it implies that the prediction error on the target sample decreases after each iteration. This result means that TRADABOOST is able to reduce the weighted training error on the source distribution, while still preserving the properties of ADABOOST on the target domain. Finally, a bound on the generalization error on the target distribution is given as follows:

$$\mathrm{R}_{\mathcal{T}}(h) \leq \mathrm{R}_{\hat{\mathcal{T}}}(h) + \mathcal{O}\left(\sqrt{\frac{N}{N_{\mathcal{T}}} VC(\mathcal{H})}\right)$$

where $VC(\mathcal{H})$ refers to the VC dimension of the considered hypothesis space $\mathcal{H}$, N is the number of iterations and $N_{\mathcal{T}}$ is the number of labeled target instances. This result is similar to the classic boosting result on a single domain [SCH 99] but bears an important difference with respect to this latter. Indeed, one can note that the assumption requiring $N_{\mathcal{T}}$ to be small may lead to overfitting of the obtained hypothesis as there is no particular term that prevents it. In practice, however, the authors report convincing classification performances that show a clear improvement with respect to no adaptation setting.

To summarize, this approach remains largely heuristic in its nature as its theoretical analysis does not provide any explicit connection between the source and target errors, does it nor relate the two domains through the discrepancy terms seen before. Consequently, it gives no particular indication of when the proposed algorithm works properly and how one can characterize the *a priori* adaptability of the two domains.

8.2.2. *SLDAB for unsupervised domain adaptation*

In another line of work, the SLDAB approach [HAB 16] considers the unsupervised domain adaptation setting where the learner takes as input a labeled source sample and an unlabeled target sample. The approach here consists of learning N weak learners from the source sample and reweighting them differently by taking into account the data from the target domain.

If the learned hypotheses $h_1, \ldots, h_N$ are the same for S and T_u, then optimizing different weighting coefficients depending on the domain to which the example belongs appears very intuitive. To this end, it seems natural (and theoretically founded) to first keep minimizing the empirical classification error on the labeled source examples with a classic boosting strategy leading to a final source model:

$$F_S^N = \sum_{t=1}^{N} \alpha_t h_t(\mathbf{x}).$$

Then, a different strategy has to be applied on the target data for which we do not have access to the labels. To this end, the authors of [HAB 16] assumed that the quality of a hypothesis h_t on the examples from T_u has to depend on its ability to minimize the proportion of margin violations of the target examples so that $|h_t(x_i)|$ is heuristically maximized. Therefore, they propose to optimize a second objective function given by a linear combination of hypotheses related to the target distribution and defined as follows:

$$F_T^N = \sum_{t=1}^{N} \beta_t h_t(\mathbf{x}),$$

where β_t depends both on the example's margin and a divergence measure between S and T_u induced by h_t. This last divergence measure is a novel idea underlying the SLDAB method where the lack of the target labels is compensated for by taking into account the divergence from the outputs of h_t between the source and the target domains. This double weighting strategy of SLDAB is illustrated on Figure 8.2.

Let us further denote the measure of divergence induced by h_t between S and T_u at each iteration t by $g_t \in [0, 1]$. The boosting scheme aims to take into account g_t in order to penalize hypotheses that do not allow the reduction of the divergence between S and T_u. This idea is similar to the divergence minimization-based approaches for domain adaptation presented previously. This objective is tackled by considering a function $f_{\mathrm{DA}} : [-1,+1] \to [-1,+1]$ such that $f_{\mathrm{DA}}(h_t(x)) = |h_t(x)| - \lambda g_t$, where $\lambda \in [0,1]$. $f_{\mathrm{DA}}(h_t(x))$ expresses the ability of h_t to not only induce large margins in an unsupervised way given by a large value for $|h_t(x)|$, but also to reduce the divergence between S and T_u given by a small value for g_n. λ plays the role of a trade-off parameter between the importance of the margin and the divergence.

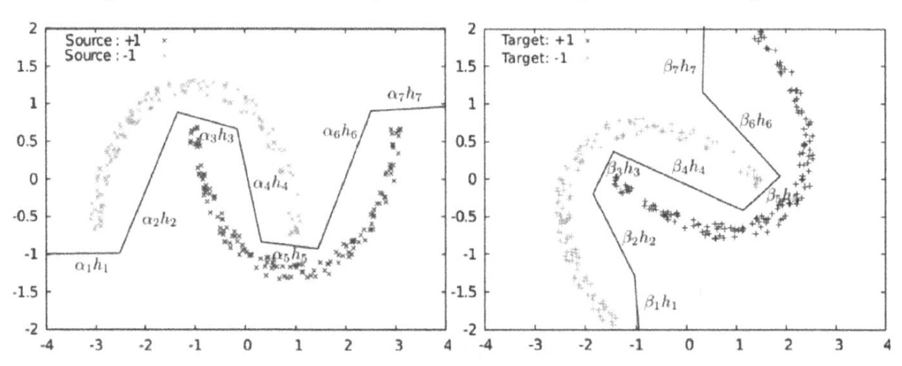

Figure 8.2. *Illustration of the intuition behind SLDAB on the Moons data set where the target data are generated after a rotation of the source examples. Source and target examples are projected in the same N-dimensional space, where the same N weak hypotheses are combined to get the final classifier. The only differences are the weights in each combination: for the source domain S (on the left), the α's parameters are optimized w.r.t. the classification error on S, while for the target distribution T (on the right), the β's parameters are learned w.r.t. the ability of the hypothesis to maximize the margins*

The boosting strategy presented here relies on a novel notion of weak learner tailored to the domain adaptation setting. Recall that the weak assumption presented in [FRE 96] states that a classifier h_n is a weak hypothesis over S if it performs at least a little bit better than random guessing, that is, $\mathrm{R}_S^{(n)}(h_t) < \frac{1}{2}$, where $\mathrm{R}_S^{(n)}(h_t)$ is the empirical error of h_t at iteration n calculated over sample S and its associated empirical source distribution $\hat{\mathcal{S}}_{\mathbf{X}}$. In SLDAB, this weak learner assumption is extended to domain adaptation in the following manner.

DEFINITION 8.4 (Weak domain adaptation learner).– *A classifier h_t learned at iteration t from a labeled source sample S drawn from $\mathcal{S}$ and an unlabeled target sample T_u drawn from $\mathcal{T}$ is a weak domain adaptation learner for T_u if $\forall \gamma \leq 1$:*

1) h_t is a weak learner for S, i.e. $\mathrm{R}_S(h_t) < \frac{1}{2}$;

2) $\hat{L}_t = \mathop{\mathbf{E}}_{x \sim \hat{\mathcal{T}}_{\mathbf{X}}^t} [|f_{DA}(h_t(x))| \leq \gamma] < \frac{\gamma}{\gamma + \max(\gamma, \lambda g_t)}$*, where* $\hat{\mathcal{T}}_{\mathbf{X}}^t$ *denotes the empirical distribution of target instances at iteration t.*

Condition (1) means that adapting $\mathcal{S}$ to $\mathcal{T}$ using a boosting scheme requires h_t to learn something new about the source labeling function at each iteration. On the other hand, condition (2) takes into account not only the ability of h_t to satisfy the margin constraint given by γ but also its capacity to reduce the divergence between S and T_u. In other words, if the divergence between S and T_u is high, the classifier is required to have a higher margin to compensate for the divergence. From definition 8.4, it turns out that:

1) if $\max(\gamma, \lambda g_t) = \gamma$, then $\frac{\gamma}{\gamma + \max(\gamma, \lambda g_t)} = \frac{1}{2}$ and condition (2) becomes a weak learner assumption over the source sample with the only difference being that $\hat{L}_t < \frac{1}{2}$ expresses a constraint in terms of margin, while $\mathrm{R}_S(h_t) < \frac{1}{2}$ stands for a bound on the classification error. Note that if this is true for any hypothesis h_t, then the divergence between the source and target distributions is rather small, and thus the underlying problem becomes more related to semi-supervised learning than to that of domain adaptation.

2) if $\max(\gamma, \lambda g_t) = \lambda g_t$, then condition (2) writes

$$\hat{L}_t < \frac{\gamma}{\gamma + \max(\gamma, \lambda g_t)} < \frac{1}{2}$$

and becomes more constraining in terms of the compensation required for a large divergence between S and T. In this case, the underlying problem corresponds to a domain adaptation problem and requires an appropriate weighting scheme.

Although our objective in this book is to focus on theoretical results and not on the algorithmic aspects of the adaptation problem, we do, however, note several important implications. First, the reader can note that condition (2) presented above makes the conditions required from a weak domain adaptation learner harder to fulfill than that of the classic weak learner. As a result, a specific algorithmic procedure must be designed in order to ensure this property which can be quite hard to verify in practice. In [HAB 16], the authors propose an approach that learns a weighted combination of k stumps algorithms (also called one-level decision trees) for which a theoretical guarantee in terms of consistency is derived because of the algorithmic robustness framework from Chapter 1. In practice, a divergence based on the perturbed variation [HAR 12] is used, but other divergences can be considered. Note that Morvant [MOR 15] related the domain disagreement $\mathrm{dis}_\rho(S, T)$ of definition 6.1 to the perturbed variation [HAR 12].

To summarize, the boosting procedure at the heart of the SLDAB method works as follows:

1) for the source labeled learning data, the updates are the same as in the classic ADABOOST algorithm;

2) for the unlabeled target instances, the weights are changed according to the margin conditions evoked previously. If a target example $\mathbf{x}$ does not match the condition $f_{\text{DA}}(h_t(\mathbf{x})) > \gamma$, then a pseudo-class $y_t = -\text{sign}(f_{\text{DA}}(h_t(\mathbf{x})))$ is assigned to $\mathbf{x}$ to simulate a misclassification. The geometrical interpretation behind this is to make the proposed approach exponentially increase the weights of the points located in an extended margin band of width $\gamma + \lambda g_t$. If $\mathbf{x}$ is outside this band, a pseudo-class $y_t = \text{sign}(f_{\text{DA}}(h_t(\mathbf{x})))$ is assigned to it which leads to an exponential decrease in the weight of the example at the next iteration. An update scheme similar to ADABOOST is then used for the target instances.

From a theoretical point of view, it can be proved that the proportion of target examples that have a margin γ decreases exponentially fast with the number of iterations. More precisely, for a given value of the divergence measure g_t obtained at each iteration, the convergence rate depends on the quantity $\max(\gamma, \lambda g_t) = \text{constant}_t \times \gamma$ and on the ratio between the number of target examples that do not satisfy the margin constraint and the number of those that satisfy it. When this quantity is small, the convergence is faster and corresponds to a situation where the margin constraint tends to be highly satisfied or when the divergence is rather small. To this end, this result has the advantage of establishing a link between the source and target distributions by means of the divergence measure. On the other hand, however, this result is rather limited as the proposed approach does not benefit from any generalization guarantees given by an upper bound on the risk with respect to the target distribution. In its turn, this implies that one cannot hope to obtain a clear indication regarding the possible performance of the target hypothesis.

8.3. Summary

In this chapter, we reviewed a different class of adaptation algorithms that rely on pseudo-labeling in order to perform adaptation between source and target domains. The presented methods follow a boosting-based strategy where a single hypothesis can be updated at each iteration or where several hypotheses produced at each iteration are combined to obtain the final predictor. As we have seen, these methods are quite different from the traditional algorithms that minimize the divergence between the marginal distributions of the source and target domains presented above, as they proceed by iteratively adapting the source classifier to the target domain where this latter is pseudo-labeled at each iteration. We presented two algorithms for both supervised and unsupervised domain adaptation settings that highlight the general intuition behind the iterative approach based on pseudo-labeling.

While iterative methods are attractive due to their simplicity and the strong intuition behind them, so far the theoretical studies of such techniques for domain adaptation remain rather limited. As a result, while they are frequently used as heuristics even in some recent methods [SAI 17], there exist no general theoretical frameworks that are able to provide a clear explanation of how to use them accurately. This leaves room for novel and interesting developments related to the theoretical guarantees for domain adaptation in these scenarios. Arguably, one can assume that providing significant theoretical results for these methods would probably mean gaining better control of the negative transfer and allowing a better understanding of iterative domain adaptation in general. Indeed, being able to efficiently (pseudo-) label target instances is *a priori* related to the development of relevant strategies that are able to control the joint errors, mentioned in Chapter 3.

Conclusions and Discussions

In this book, we presented an overview of existing theoretical guarantees that were proved for the domain adaptation problem, a learning setting that extends traditional learning paradigms to the case where the model is learned and deployed on samples coming from different yet related probability distributions. The cited theoretical results often took a shape of generalization bounds where the goal is to relate the error of a model on the training (also called source) domain to that of the test (also called target) domain. To this end, we note that the state-of-the-art results presented are highly intuitive as they manage to explicitly introduce the dependence of the relationship between the two errors mentioned above to the similarity of their data-generating probability distributions and that of their corresponding labeling functions. Consequently, this two-way similarity between the source and target domains characterizes both unsupervised proximity of two domains by comparing their marginal distributions, and the possible labelings of their samples by looking for a good model that has a low error rate with respect to them. This general trade-off is preserved, in one way or another, in the majority of published results on the subject and thus can be considered as a cornerstone of the modern domain adaptation theory.

Starting from this general philosophy, many variations of generalization bounds for domain adaptation were proposed. Some of them aimed at introducing novel divergence measures, such as the discrepancy distance and the integral probability metrics, in order to tackle the drawbacks of the previously used $\mathcal{H}\Delta\mathcal{H}$ divergence. Others introduced a different way of approaching the problem by considering the weighted majority vote point of view in the PAC–Bayesian framework. In addition to this, the literature on the domain adaptation theory also presents contributions for those cases where one only has access to the hypothesis produced based on the observed source data sample, without having this latter during the adaptation step. Finally, several contributions showing the utility of boosting-based procedures where also proposed and analyzed in the domain adaptation setting. This, together with the impossibility results highlighting cases and conditions when adaptation is not possible or intractable, gives a large body of work that is covered in this book.

As would any survey that gives an overview of a certain scientific field, this book would be incomplete without identifying those problems that remain open. In the context of domain adaptation theory, these open problems can arguably be split into two main categories where the first one is related to the domain adaptation problem itself and the second is related to other learning scenarios similar to domain adaptation. For the first category, one important open problem is that of characterizing the *a priori* adaptability of the adaptation given by the joint error term. Indeed, this term is often assumed to be small for domain adaptation to be possible even though so far no consistent algorithms have been proposed to estimate it from a handful of labeled target data. On the other hand, domain adaptation has recently been extended to the open-set setting where both source and target domains are allowed to have non-overlapping classes. To the best of our knowledge, there are no theoretical results that analyze this generalized adaptation setting. This latter point brings us to the second category of open problems related to learning scenarios similar to that of domain adaptation. Among them, we can primarily highlight zero- and one-shot learning problems where one has to learn from a sample that contains no examples, or only one, of certain classes appearing in the test data. Obviously, both these problems are tightly related to the domain adaptation problem and naturally inherit the theoretical guarantees reviewed in this book. However, it remains unclear whether one can provide a theoretical analysis of these two settings without casting them as an adaptation problem. The potential interest of doing so would be to obtain generalization bounds that are proper to zero- and one-shot learning in order to reveal their own peculiarities and insights into their differences with respect to the traditional domain adaptation setting.

Finally, we note that this book has not covered areas such as multitask learning, learning-to-learn and lifelong learning, even though they are all tightly related to transfer learning in general and domain adaptation in particular. This particular choice was made for the sake of clarity and in order to remain focused on one particular problem that is vast enough on its own. We also admit that there are certainly other relevant papers providing guarantees for domain adaptation that were not included in this book. This field, however, is so large and recent advances are published at such a great pace that we are simply not able to keep up with it and to report all possible results without breaking the general structure and narrative of our survey.

Appendices

Appendix 1

Proofs of the Main Results of Chapter 3

THEOREM 3.1 ([BEN 07]).– *Given two domains $\mathcal{S}$ and $\mathcal{T}$ over $\mathbf{X} \times Y$ and a hypothesis class $\mathcal{H}$, the following holds for any $h \in \mathcal{H}$*

$$\forall h \in \mathcal{H}, \quad \mathrm{R}_{\mathcal{T}}^{\ell_{01}}(h) \leq \mathrm{R}_{\mathcal{S}}^{\ell_{01}}(h) + d_1(\mathcal{S}_{\mathbf{X}}, \mathcal{T}_{\mathbf{X}})$$

$$+ \min \left\{ \mathop{\mathbf{E}}_{\mathbf{x} \sim \mathcal{S}_{\mathbf{X}}} [|f_{\mathcal{S}}(\mathbf{x}) - f_{\mathcal{T}}(\mathbf{x})|], \mathop{\mathbf{E}}_{\mathbf{x} \sim \mathcal{T}_{\mathbf{X}}} [|f_{\mathcal{T}}(\mathbf{x}) - f_{\mathcal{S}}(\mathbf{x})|] \right\},$$

where $f_{\mathcal{S}}(\mathbf{x})$ and $f_{\mathcal{T}}(\mathbf{x})$ are source and target true labeling functions associated with $\mathcal{S}$ and $\mathcal{T}$, respectively.

PROOF.– Let $\phi_{\mathcal{S}}(\mathbf{x})$ and $\phi_{\mathcal{T}}(\mathbf{x})$ be density functions of $\mathcal{S}_{\mathbf{X}}$ and $\mathcal{T}_{\mathbf{X}}$, respectively. Then, we have:

$$\mathrm{R}_{\mathcal{T}}^{\ell_{01}}(h) = \mathrm{R}_{\mathcal{T}}^{\ell_{01}}(h) - \mathrm{R}_{\mathcal{S}}^{\ell_{01}}(h) + \mathrm{R}_{\mathcal{S}}^{\ell_{01}}(h) + \mathrm{R}_{\mathcal{S}}^{\ell_{01}}(h, f_{\mathcal{T}}) - \mathrm{R}_{\mathcal{S}}^{\ell_{01}}(h, f_{\mathcal{T}})$$

$$\leq \mathrm{R}_{\mathcal{S}}^{\ell_{01}}(h) + |\mathrm{R}_{\mathcal{S}}^{\ell_{01}}(h, f_{\mathcal{T}}) - \mathrm{R}_{\mathcal{S}}^{\ell_{01}}(h, f_{\mathcal{S}})| + |\mathrm{R}_{\mathcal{T}}^{\ell_{01}}(h, f_{\mathcal{T}}) - \mathrm{R}_{\mathcal{S}}^{\ell_{01}}(h, f_{\mathcal{T}})|$$

$$\leq \mathrm{R}_{\mathcal{S}}^{\ell_{01}}(h) + \mathop{\mathbf{E}}_{\mathbf{x} \sim \mathcal{S}_{\mathbf{X}}} [|f_{\mathcal{S}}(\mathbf{x}) - f_{\mathcal{T}}(\mathbf{x})|] + \int |\phi_{\mathcal{S}}(\mathbf{x}) - \phi_{\mathcal{T}}(\mathbf{x})||h(\mathbf{x}) - f_{\mathcal{T}}(\mathbf{x})| dx$$

$$\leq \mathrm{R}_{\mathcal{S}}^{\ell_{01}}(h) + \mathop{\mathbf{E}}_{\mathbf{x} \sim \mathcal{S}_{\mathbf{X}}} [|f_{\mathcal{S}}(\mathbf{x}) - f_{\mathcal{T}}(\mathbf{x})|] + d_1(\mathcal{S}_{\mathbf{X}}, \mathcal{T}_{\mathbf{X}}).$$

Note that in the first line of the proof, one could have added and subtracted the error of the target domain, which would result in the same inequality but with the expectation taken over $\mathcal{T}_{\mathbf{X}}$. The final result is thus obtained by choosing the smaller of the two. □

LEMMA 3.2 ([BEN 10a]).– *Let $\mathcal{H}$ be a hypothesis space. Then, for two unlabeled samples S_u, T_u of size m we have*

$$\hat{d}_{\mathcal{H}\Delta\mathcal{H}}(S_u, T_u) = 2\left(1 - \min_{h \in \mathcal{H}\Delta\mathcal{H}}\left[\frac{1}{m}\sum_{\mathbf{x}:h(\mathbf{x})=0}\mathbf{I}\left[\mathbf{x} \in S_u\right] + \frac{1}{m}\sum_{\mathbf{x}:h(\mathbf{x})=1}\mathbf{I}\left[\mathbf{x} \in T_u\right]\right]\right).$$

PROOF.– In order to proceed, the authors show that for any hypothesis h and corresponding set $I(h)$ of positively labeled instances,

$$1 - \left[\frac{1}{m}\sum_{\mathbf{x}:h(\mathbf{x})=0}\mathbf{I}\left[\mathbf{x} \in S_u\right] + \frac{1}{m}\sum_{\mathbf{x}:h(x)=1}\mathbf{I}\left[\mathbf{x} \in T_u\right]\right] = \Pr_{S_u} I(h) - \Pr_{T_u} I(h).$$

This result is obtained as follows:

$$\begin{aligned}
&1 - \left[\frac{1}{m}\sum_{\mathbf{x}:h(\mathbf{x})=0}\mathbf{I}\left[\mathbf{x} \in S_u\right] + \frac{1}{m}\sum_{\mathbf{x}:h(x)=1}\mathbf{I}\left[\mathbf{x} \in T_u\right]\right] \\
&= \frac{1}{2m}\sum_{\mathbf{x}:h(\mathbf{x})=0}\left(\mathbf{I}\left[\mathbf{x} \in S_u\right] + \mathbf{I}\left[\mathbf{x} \in T_u\right]\right) + \frac{1}{2m}\sum_{\mathbf{x}:h(\mathbf{x})=1}\left(\mathbf{I}\left[\mathbf{x} \in S_u\right] + \mathbf{I}\left[\mathbf{x} \in T_u\right]\right) \\
&\quad - \frac{1}{m}\left(\sum_{\mathbf{x}:h(\mathbf{x})=0}\mathbf{I}\left[\mathbf{x} \in S_u\right] + \sum_{\mathbf{x}:h(\mathbf{x})=1}\mathbf{I}\left[\mathbf{x} \in T_u\right]\right) \\
&= \frac{1}{2m}\sum_{\mathbf{x}:h(\mathbf{x})=0}\left(\mathbf{I}\left[\mathbf{x} \in S_u\right] - \mathbf{I}\left[\mathbf{x} \in T_u\right]\right) - \frac{1}{2m}\sum_{\mathbf{x}:h(\mathbf{x})=1}\left(\mathbf{I}\left[\mathbf{x} \in S_u\right] - \mathbf{I}\left[\mathbf{x} \in T_u\right]\right) \\
&= \frac{1}{2}\left(1 - \Pr_{S_u} I(h) - (1 - \Pr_{T_u} I(h))\right) + \frac{1}{2}\left(\Pr_{T_u} I(h) - \Pr_{S_u} I(h))\right) \\
&= \Pr_{S_u} I(h) - \Pr_{T_u} I(h).
\end{aligned}$$

□

LEMMA 3.3 ([BEN 10a]).– *Let $\mathcal{S}$ and $\mathcal{T}$ be two domains on $\mathbf{X} \times Y$. For any pair of hypotheses $(h, h') \in \mathcal{H}\Delta\mathcal{H}^2$, we have*

$$\left|\mathrm{R}_{\mathcal{T}}^{\ell_{01}}(h, h') - \mathrm{R}_{\mathcal{S}}^{\ell_{01}}(h, h')\right| \leq \frac{1}{2} d_{\mathcal{H}\Delta\mathcal{H}}(\mathcal{S}_{\mathbf{X}}, \mathcal{T}_{\mathbf{X}}).$$

PROOF.– From the definition of the $\mathcal{H}\Delta\mathcal{H}$-divergence, we get:

$$d_{\mathcal{H}\Delta\mathcal{H}}(\mathcal{S}_{\mathbf{X}}, \mathcal{T}_{\mathbf{X}}) = 2 \sup_{h,h' \in \mathcal{H}} \left| \mathbf{Pr}_{\mathbf{x}\sim\mathcal{S}_{\mathbf{X}}} (h(\mathbf{x}) \neq h(\mathbf{x}')) - \mathbf{Pr}_{\mathbf{x}\sim\mathcal{T}_{\mathbf{X}}} (h(\mathbf{x}) \neq h(\mathbf{x}')) \right|$$

$$= 2 \sup_{h,h' \in \mathcal{H}} \left| \mathrm{R}_{\mathcal{S}}^{\ell_{01}}(h, h') - \mathrm{R}_{\mathcal{T}}^{\ell_{01}}(h, h') \right| \geq \frac{1}{2} d_{\mathcal{H}\Delta\mathcal{H}}(\mathcal{S}_{\mathbf{X}}, \mathcal{T}_{\mathbf{X}}).$$

□

THEOREM 3.2 ([BEN 10a]).– *Let $\mathcal{H}$ be a hypothesis space of VC dimension $VC(\mathcal{H})$. If S_u, T_u are unlabeled samples of size m' each, drawn independently from $\mathcal{S}_{\mathbf{X}}$ and $\mathcal{T}_{\mathbf{X}}$, respectively, then for any $\delta \in (0, 1)$ with probability at least $1-\delta$ over the random choice of the samples, we have that for all $h \in \mathcal{H}$*

$$\mathrm{R}_{\mathcal{T}}^{\ell_{01}}(h) \leq \mathrm{R}_{\mathcal{S}}^{\ell_{01}}(h) + \tfrac{1}{2}\hat{d}_{\mathcal{H}\Delta\mathcal{H}}(S_u, T_u) + 4\sqrt{\frac{2\,VC(\mathcal{H})\log(2m') + \log(\frac{2}{\delta})}{m'}} + \lambda,$$

where λ is the combined error of the ideal hypothesis h^* that minimizes $\mathrm{R}_{\mathcal{S}}(h) + \mathrm{R}_{\mathcal{T}}(h)$.

PROOF.–

$$\mathrm{R}_{\mathcal{T}}^{\ell_{01}}(h) \leq \mathrm{R}_{\mathcal{T}}^{\ell_{01}}(h^*) + \mathrm{R}_{\mathcal{T}}^{\ell_{01}}(h, h^*) \quad \text{[A1.1]}$$

$$\leq \mathrm{R}_{\mathcal{T}}^{\ell_{01}}(h^*) + \mathrm{R}_{\mathcal{S}}^{\ell_{01}}(h, h^*) + |\mathrm{R}_{\mathcal{T}}^{\ell_{01}}(h, h^*) - \mathrm{R}_{\mathcal{S}}^{\ell_{01}}(h, h^*)| \quad \text{[A1.2]}$$

$$\leq \mathrm{R}_{\mathcal{T}}^{\ell_{01}}(h^*) + \mathrm{R}_{\mathcal{S}}^{\ell_{01}}(h) + \mathrm{R}_{\mathcal{S}}^{\ell_{01}}(h^*) + \tfrac{1}{2} d_{\mathcal{H}\Delta\mathcal{H}}(\mathcal{S}_{\mathbf{X}}, \mathcal{T}_{\mathbf{X}}) \quad \text{[A1.3]}$$

$$= \mathrm{R}_{\mathcal{S}}^{\ell_{01}}(h) + \tfrac{1}{2} d_{\mathcal{H}\Delta\mathcal{H}}(\mathcal{S}_{\mathbf{X}}, \mathcal{T}_{\mathbf{X}}) + \lambda \quad \text{[A1.4]}$$

$$\leq \mathrm{R}_{\mathcal{S}}^{\ell_{01}}(h) + \tfrac{1}{2}\hat{d}_{\mathcal{H}\Delta\mathcal{H}}(S_u, T_u) + 4\sqrt{\frac{2\,VC(\mathcal{H})\log(2m') + \log(\frac{2}{\delta})}{m'}} + \lambda. \quad \text{[A1.5]}$$

The second and fourth lines are obtained using the triangular inequality applied to the error function. Here, the third line is obtained using lemma 3.3, the fourth line is due to the definition of the λ term. The final result is obtained by applying lemma 3.1. □

THEOREM 3.3 ([BLI 08, BEN 10a]).– *Let $\mathcal{H}$ be a hypothesis space of VC dimension $VC(\mathcal{H})$. Let $\mathcal{S}$ and $\mathcal{T}$ be the source and target domain, respectively, defined on $\mathbf{X} \times Y$. Let S_u, T_u be unlabeled samples of size m' each, drawn independently from $\mathcal{S}_{\mathbf{X}}$ and $\mathcal{T}_{\mathbf{X}}$, respectively. Let S be a labeled sample of size m generated by drawing $\beta\, m$ points from $\mathcal{T}$ ($\beta \in [0, 1]$) and $(1-\beta)\, m$ points from $\mathcal{S}$ and labeling them according* *to f_S and f_T, respectively. If $\hat{h} \in \mathcal{H}$ is the empirical minimizer of $\hat{\mathrm{R}}^{\alpha}(h)$ on S and $h^*_T = \operatorname*{argmin}_{h\in\mathcal{H}} \mathrm{R}^{\ell_{01}}_{\mathcal{T}}(h)$, then for any $\delta \in (0,1)$, with probability at least $1-\delta$ over the random choice of the samples, we have*

$$\mathrm{R}^{\ell_{01}}_{\mathcal{T}}(\hat{h}) \;\le\; \mathrm{R}^{\ell_{01}}_{\mathcal{T}}(h^*_T) + c_1 + c_2,$$

where

$$\begin{aligned} c_1 &= 4\sqrt{\frac{\alpha^2}{\beta} + \frac{(1-\alpha)^2}{1-\beta}}\sqrt{\frac{2\,VC(\mathcal{H})\log(2(m+1)) + 2\log(\frac{8}{\delta})}{m}}, \\ \textit{and } c_2 &= 2(1-\alpha)\left(\frac{1}{2}d_{\mathcal{H}\Delta\mathcal{H}}(S_u,T_u) + 4\sqrt{\frac{2\,VC(\mathcal{H})\log(2m') + \log(\frac{8}{\delta})}{m'}} + \lambda\right). \end{aligned} \qquad [3.2]$$

PROOF.– In order to prove this theorem, one first needs to establish a concentration inequality for the true and empirical combined error functions. To this end, let us consider $X_1, \ldots, X_{\beta m} \in [0, \frac{\alpha}{\beta}]$ be random variables having values $\frac{\alpha}{\beta}|h(\mathbf{x}) - f_{\mathcal{T}}(\mathbf{x})|$ for the βm instances $\mathbf{x} \in S_{\mathcal{T}}$. Similarly, let $X_{\beta m+1}, \ldots, X_m \in [0, \frac{1-\alpha}{1-\beta}]$ be variables that take on values $\frac{1-\alpha}{1-\beta}|h(\mathbf{x}) - f_{\mathcal{T}}(\mathbf{x})|$ for the βm instances $\mathbf{x} \in S_{\mathcal{S}}$.

Then, we can rewrite the empirical combined error as follows:

$$\begin{aligned} \hat{\mathrm{R}}^{\alpha}(h) &= \alpha \mathrm{R}^{\ell_{01}}_{\hat{\mathcal{T}}}(h) + (1-\alpha)\mathrm{R}^{\ell_{01}}_{\hat{\mathcal{S}}}(h) \\ &= \frac{\alpha}{\beta m}\sum_{\mathbf{x}\in S_{\mathcal{T}}} |h(\mathbf{x}) - f_{\mathcal{T}}(\mathbf{x})| + \frac{(1-\alpha)}{(1-\beta)m}\sum_{\mathbf{x}\in S_{\mathcal{S}}} |h(\mathbf{x}) - f_{\mathcal{S}}(\mathbf{x})| = \sum_{i=1}^{m} X_i. \end{aligned}$$

Furthermore, we can show that, due to the linearity of expectation:

$$\begin{aligned} \mathbf{E}\left[\mathrm{R}^{\alpha}(h)\right] &= \frac{1}{m}\left(\beta m \frac{\alpha}{\beta}\mathrm{R}^{\ell_{01}}_{\mathcal{T}}(h) + (1-\beta)m\frac{1-\alpha}{1-\beta}\mathrm{R}^{\ell_{01}}_{\mathcal{S}}(h)\right) \\ &= \alpha\mathrm{R}^{\ell_{01}}_{\mathcal{T}}(h) + (1-\alpha)\mathrm{R}^{\ell_{01}}_{\mathcal{S}}(h). \end{aligned}$$

Using these two equality's, we can apply the Hoeffding inequality [HOE 63] to obtain the desired concentration results:

$$\begin{aligned} \mathbf{Pr}\left\{|\hat{\mathrm{R}}^{\alpha}(h) - \mathrm{R}^{\alpha}(h)| > \varepsilon\right\} &\le 2\exp\left(\frac{-2m\epsilon^2}{\sum_{i=1}^{m}\mathrm{range}^2(X_i)}\right) \\ &\le 2\exp\left(\frac{-2m\epsilon^2}{\sum_{i=1}^{m}\beta m(\frac{\alpha}{\beta})^2 + (1-\beta)m(\frac{1-\alpha}{1-\beta})^2}\right) \end{aligned}$$

$$\leq 2\exp\left(\frac{-2m\epsilon^2}{\sum_{i=1}^{m}\frac{\alpha^2}{\beta}+\frac{(1-\alpha)^2}{1-\beta}}\right).$$

We can now proceed to the proof of the main theorem.

$$\mathrm{R}_{\mathcal{T}}^{\ell_{01}}(\hat{h}) \leq \mathrm{R}^{\alpha}(\hat{h}) + (1-\alpha)(\tfrac{1}{2}d_{\mathcal{H}\Delta\mathcal{H}}(\mathcal{S}_{\mathbf{X}},\mathcal{T}_{\mathbf{X}})+\lambda) \qquad [A1.6]$$

$$\leq \hat{\mathrm{R}}^{\alpha}(\hat{h}) + 2\sqrt{\frac{\alpha^2}{\beta}+\frac{(1-\alpha)^2}{1-\beta}}\sqrt{\frac{2\,VC(\mathcal{H})\log(2(m+1))+2\log(\frac{8}{\delta})}{m}} \qquad [A1.7]$$

$$+ (1-\alpha)(\tfrac{1}{2}d_{\mathcal{H}\Delta\mathcal{H}}(\mathcal{S}_{\mathbf{X}},\mathcal{T}_{\mathbf{X}})+\lambda)$$

$$\leq \hat{\mathrm{R}}^{\alpha}(h_{\mathcal{T}}^{*}) + 2\sqrt{\frac{\alpha^2}{\beta}+\frac{(1-\alpha)^2}{1-\beta}}\sqrt{\frac{2\,VC(\mathcal{H})\log(2(m+1))+2\log(\frac{8}{\delta})}{m}} \qquad [A1.8]$$

$$+ (1-\alpha)(\tfrac{1}{2}d_{\mathcal{H}\Delta\mathcal{H}}(\mathcal{S}_{\mathbf{X}},\mathcal{T}_{\mathbf{X}})+\lambda)$$

$$\leq \hat{\mathrm{R}}^{\alpha}(h_{\mathcal{T}}^{*}) + 4\sqrt{\frac{\alpha^2}{\beta}+\frac{(1-\alpha)^2}{1-\beta}}\sqrt{\frac{2\,VC(\mathcal{H})\log(2(m+1))+2\log(\frac{8}{\delta})}{m}} \qquad [A1.9]$$

$$+ (1-\alpha)(\tfrac{1}{2}d_{\mathcal{H}\Delta\mathcal{H}}(\mathcal{S}_{\mathbf{X}},\mathcal{T}_{\mathbf{X}})+\lambda)$$

$$\leq \mathrm{R}_{\mathcal{T}}^{\ell_{01}}(h_{\mathcal{T}}^{*}) + 4\sqrt{\frac{\alpha^2}{\beta}+\frac{(1-\alpha)^2}{1-\beta}}\sqrt{\frac{2\,VC(\mathcal{H})\log(2(m+1))+2\log(\frac{8}{\delta})}{m}} \qquad [A1.10]$$

$$+ 2(1-\alpha)(\tfrac{1}{2}d_{\mathcal{H}\Delta\mathcal{H}}(\mathcal{S}_{\mathbf{X}},\mathcal{T}_{\mathbf{X}})+\lambda)$$

$$\leq \mathrm{R}_{\mathcal{T}}^{\ell_{01}}(h_{\mathcal{T}}^{*}) + 4\sqrt{\frac{\alpha^2}{\beta}+\frac{(1-\alpha)^2}{1-\beta}}\sqrt{\frac{2\,VC(\mathcal{H})\log(2(m+1))+2\log(\frac{8}{\delta})}{m}} \qquad [A1.11]$$

$$+ 2(1-\alpha)(\frac{1}{2}d_{\mathcal{H}\Delta\mathcal{H}}(S_u,T_u) + 4\sqrt{\frac{2\,VC(\mathcal{H})\log(2m')+\log(\frac{8}{\delta})}{m'}}+\lambda).$$

The proof follows the standard theory of uniform convergence for empirical risk minimizers where lines 1 and 5 are obtained by observing that

$$|\mathrm{R}^{\alpha}(h) - \mathrm{R}_{\mathcal{T}}^{\ell_{01}}(h)| = |\alpha\mathrm{R}_{\mathcal{T}}^{\ell_{01}}(h) + (1-\alpha)\mathrm{R}_{\mathcal{S}}^{\ell_{01}}(h) - \mathrm{R}_{\mathcal{T}}^{\ell_{01}}(h)|$$
$$= |(1-\alpha)(\mathrm{R}_{\mathcal{S}}^{\ell_{01}}(h) - \mathrm{R}_{\mathcal{T}}^{\ell_{01}}(h))| \leq (1-\alpha)(d_{\mathcal{H}\Delta\mathcal{H}}(\mathcal{S}_{\mathbf{X}},\mathcal{T}_{\mathbf{X}})+\lambda).$$

Line 3 follows from the definition of $\hat{h}$ and h_T^* and line 6 is a consequence of lemma 3.1. Finally, lines 2 and 4 are obtained based on the concentration inequality obtained for $\mathrm{R}_{\hat{\mathcal{S}}}^{\alpha}(h)$. □

PROPOSITION 3.1 ([MAN 09a]).– *Given two domains $\mathcal{S}$ and $\mathcal{T}$ over $\mathbf{X} \times Y$, let $\mathcal{H}$ be a hypothesis class and let $\ell : Y \times Y \to \mathbb{R}_+$ define a loss function that is bounded, $\forall (y, y') \in Y^2$, $\ell(y, y') \leq M$ for some $M > 0$. Then, for any hypothesis $h \in \mathcal{H}$, we have*

$$disc_\ell(\mathcal{S}_\mathbf{X}, \mathcal{T}_\mathbf{X}) \leq M\, d_1(\mathcal{S}_\mathbf{X}, \mathcal{T}_\mathbf{X}).$$

PROOF.– For any distributions $\mathcal{S}_\mathbf{X}$ and $\mathcal{T}_\mathbf{X}$ that are absolutely continuous with respect to their density functions $\phi_\mathcal{S}$ and $\phi_\mathcal{T}$, the following holds:

$$\begin{aligned} disc_\ell(\mathcal{S}_\mathbf{X}, \mathcal{T}_\mathbf{X}) &= \max_{h,h' \in \mathcal{H}} \left| \mathop{\mathbf{E}}_{\mathbf{x} \sim \mathcal{S}_\mathbf{X}} [\ell(h'(\mathbf{x}), h(\mathbf{x}))] - \mathop{\mathbf{E}}_{\mathbf{x} \sim \mathcal{T}_\mathbf{X}} [\ell(h'(\mathbf{x}), h(\mathbf{x}))] \right| \\ &= \max_{h,h' \in \mathcal{H}} \left| \int_\mathbf{X} (\phi_\mathcal{S} - \phi_\mathcal{T}) \ell(h'(\mathbf{x}), h(\mathbf{x}))\, dx \right| \\ &\leq \max_{h,h' \in \mathcal{H}} \int_\mathbf{X} |(\phi_\mathcal{S} - \phi_\mathcal{T}) \ell(h'(\mathbf{x}), h(\mathbf{x}))|\, dx \\ &\leq M \int_\mathbf{X} |\phi_\mathcal{S} - \phi_\mathcal{T})|\, dx = M d_1(\mathcal{S}_\mathbf{X}, \mathcal{T}_\mathbf{X}). \end{aligned}$$

□

LEMMA 3.4 ([MAN 09a]).– *Let $\mathcal{H}$ be a hypothesis class, and let $\ell : Y \times Y \to \mathbb{R}_+$ define a loss function that is bounded, $\forall (y, y') \in Y^2$, $\ell(y, y') \leq M$ for some $M > 0$ and $L_\mathcal{H} = \{\mathbf{x} \to \ell(h'(\mathbf{x}), h(\mathbf{x})) : h, h' \in \mathcal{H}\}$. Let $\mathcal{D}_\mathbf{X}$ be a distribution over $\mathbf{X}$ and $\hat{\mathcal{D}}_\mathbf{X}$ denote the corresponding empirical distribution for a sample $S = (\mathbf{x}_1, \ldots, \mathbf{x}_m)$. Then, for any $\delta \in (0, 1)$, with probability at least $1 - \delta$ over the choice of sample S, we have*

$$disc_\ell(\mathcal{D}_\mathbf{X}, \hat{\mathcal{D}}_\mathbf{X}) \leq \mathcal{R}_S(L_\mathcal{H}) + 3M\sqrt{\frac{\log \frac{2}{\delta}}{2m}},$$

where $\mathcal{R}_S(L_\mathcal{H})$ is the empirical Rademacher complexity of $L_\mathcal{H}$ based on the observations from S.

PROOF.– To proceed, we scale the loss function ℓ to the $[0, 1]$ interval by dividing it by M, and denote the new considered hypothesis class by $L_\mathcal{H}/M$. By theorem

1.3 applied to $L_{\mathcal{H}}/M$, for any $\delta > 0$, with probability at least $1 - \delta$, the following inequality holds for all $h, h' \in \mathcal{H}$:

$$\frac{\mathrm{R}^{\ell}_{\mathcal{D}_{\mathbf{X}}}(h, h')}{M} \leq \frac{\mathrm{R}^{\ell}_{\hat{\mathcal{D}}_{\mathbf{X}}}(h, h')}{M} + \mathcal{R}_S(L_{\mathcal{H}}/M) + 3\sqrt{\frac{\ln\frac{2}{\delta}}{2m}}$$

The empirical Rademacher complexity has the property that $\mathcal{R}_S(\alpha\mathcal{H}) = \alpha\mathcal{R}_S(\mathcal{H})$ for any hypothesis class $\mathcal{H}$ and positive real number α. Consequently, $\mathcal{R}_S(L_{\mathcal{H}}/M) = \frac{1}{M}\mathcal{R}_S(L_{\mathcal{H}})$, which established the final result of the proposition. □

COROLLARY 3.1 ([MAN 09a]).– *Let $\mathcal{S}$ and $\mathcal{T}$ be the source and the target domain over $\mathbf{X} \times Y$, respectively. Let $\mathcal{H}$ be a hypothesis class and $\ell_q : Y \times Y \to \mathbb{R}_+$ be a loss function that is bounded, $\forall (y, y') \in Y^2$, $\ell_q(y, y') \leq M$ for some $M > 0$, and defined as $\forall (y, y') \in Y^2$, $\ell_q(y, y') = |y - y'|^q$ for some q. Let S_u and T_u be samples of size m_s and m_t drawn independently from $\mathcal{S}_{\mathbf{X}}$ and $\mathcal{T}_{\mathbf{X}}$. Denote by $\hat{\mathcal{S}}_{\mathbf{X}}, \hat{\mathcal{T}}_{\mathbf{X}}$ the empirical distributions corresponding to $\mathcal{S}_{\mathbf{X}}$ and $\mathcal{T}_{\mathbf{X}}$. Then, for any $\delta \in (0, 1)$, with probability at least $1 - \delta$ over the random choice of samples, we have*

$$disc_{\ell_q}(\mathcal{S}_{\mathbf{X}}, \mathcal{T}_{\mathbf{X}}) \leq disc_{\ell_q}(\hat{\mathcal{S}}_{\mathbf{X}}, \hat{\mathcal{T}}_{\mathbf{X}}) + 4q\left(\mathcal{R}_{S_u}(\mathcal{H}) + \mathcal{R}_{T_u}(\mathcal{H})\right)$$
$$+3M\left(\sqrt{\frac{\log(\frac{4}{\delta})}{2m_s}} + \sqrt{\frac{\log(\frac{4}{\delta})}{2m_t}}\right).$$

PROOF.– To present the proof, we first note that the function $f : \mathbf{x} \to \mathbf{x}^q$ is q-Lipschitz for $x \in [0, 1]$:

$$|f(x') - f(x)| \leq q|x' - x|,$$

and $f(0) = 0$. In the considered case when $\ell = \ell_q$, the underlying space of losses is then $L_{\mathcal{H}} = \{\mathbf{x} \to |(h'(\mathbf{x}) - h(\mathbf{x}))|^q : h, h' \in \mathcal{H}\}$. Thus, by Talagrand's contraction lemma [LED 91], $\mathcal{R}_S(L_{\mathcal{H}})$ is bounded by $2q\mathcal{R}_S(\mathcal{H}')$ with $\mathcal{H}' = \{\mathbf{x} \to (h'(x) - h(x)) : h, h' \in \mathcal{H}\}$.

Then, $\mathcal{R}_S(\mathcal{H}')$ can be written and bounded as follows:

$$\mathcal{R}_S(\mathcal{H}') = \mathop{\mathbf{E}}_{\sigma}[\sup_{h,h'} \frac{1}{m}|\sigma_i(h(\mathbf{x}_i) - h'(\mathbf{x}_i))|]$$
$$\leq \mathop{\mathbf{E}}_{\sigma}[\sup_{h} \frac{1}{m}|\sigma_i h(\mathbf{x}_i)| + \mathop{\mathbf{E}}_{\sigma}[\sup_{h'} \frac{1}{m}|\sigma_i(h(\mathbf{x}_i)| = 2\mathcal{R}_S(\mathcal{H}).$$

From this, we get that:

$$disc_{\ell_q}(\mathcal{D}_{\mathbf{X}}, \hat{\mathcal{D}}_{\mathbf{X}}) \leq 4q\mathcal{R}_S(\mathcal{H}) + 3M\sqrt{\frac{\log \frac{2}{\delta}}{2m}}. \quad [A1.12]$$

The final results are obtained by using the triangle inequality twice such that:

$$disc_{\ell_q}(\mathcal{S}_{\mathbf{X}}, \mathcal{T}_{\mathbf{X}}) \leq disc_{\ell_q}(\hat{\mathcal{S}}_{\mathbf{X}}, \mathcal{S}_{\mathbf{X}}) + disc_{\ell_q}(\hat{\mathcal{T}}_{\mathbf{X}}, \mathcal{T}_{\mathbf{X}}) + disc_{\ell_q}(\hat{\mathcal{S}}_{\mathbf{X}}, \hat{\mathcal{T}}_{\mathbf{X}})$$

and by applying the inequality in (A1.12) to $disc_{\ell_q}(\hat{\mathcal{S}}_{\mathbf{X}}, \mathcal{S}_{\mathbf{X}})$ and $disc_{\ell_q}(\hat{\mathcal{T}}_{\mathbf{X}}, \mathcal{T}_{\mathbf{X}})$. □

THEOREM 3.4 ([MAN 09a]).– *Let $\mathcal{S}$ and $\mathcal{T}$ be the source and the target domain over $\mathbf{X} \times Y$, respectively. Let $\mathcal{H}$ be a hypothesis class; $\ell : Y \times Y \to \mathbb{R}_+$ be a loss function that is symmetric, obeys the triangle inequality and is bounded; $\forall (y, y') \in Y^2$; and $\ell(y, y') \leq M$ for some $M > 0$. Then, for $h^*_{\mathcal{S}} = \underset{h \in \mathcal{H}}{\text{argmin}}\ \mathrm{R}^{\ell}_{\mathcal{S}}(h)$ and $h^*_{\mathcal{T}} = \underset{h \in \mathcal{H}}{\text{argmin}}\ \mathrm{R}^{\ell}_{\mathcal{T}}(h)$ denoting the ideal hypotheses for the source and target domains, we have*

$$\forall h \in \mathcal{H},\ \mathrm{R}^{\ell}_{\mathcal{T}}(h) \leq \mathrm{R}^{\ell}_{\mathcal{S}}(h, h^*_{\mathcal{S}}) + disc_{\ell}(\mathcal{S}_{\mathbf{X}}, \mathcal{T}_{\mathbf{X}}) + \epsilon,$$

$$\text{where } \mathrm{R}^{\ell}_{\mathcal{S}}(h, h^*_{\mathcal{S}}) = \underset{\mathbf{x} \sim \mathcal{S}_{\mathbf{X}}}{\mathbf{E}}\ \ell\left(h(\mathbf{x}), h^*_{\mathcal{S}}(\mathbf{x})\right) \text{ and } \epsilon = \mathrm{R}^{\ell}_{\mathcal{T}}(h^*_{\mathcal{T}}) + \mathrm{R}^{\ell}_{\mathcal{S}}(h^*_{\mathcal{T}}, h^*_{\mathcal{S}}).$$

PROOF.– Fix some $h \in \mathcal{H}$. Applying the triangle inequality to ℓ and using the definition of $disc_{\ell}$ gives the following result:

$$\begin{aligned}\mathrm{R}^{\ell}_{\mathcal{T}}(h) &\leq \mathrm{R}^{\ell}_{\mathcal{T}}(h, h^*_{\mathcal{S}}) + \mathrm{R}^{\ell}_{\mathcal{T}}(h^*_{\mathcal{S}}, h^*_{\mathcal{T}}) + \mathrm{R}^{\ell}_{\mathcal{T}}(h^*_{\mathcal{T}}, f_{\mathcal{T}}) \\ &\leq \mathrm{R}^{\ell}_{\mathcal{S}}(h, h^*_{\mathcal{S}}) + disc_{\ell}(\mathcal{S}_{\mathbf{X}}, \mathcal{T}_{\mathbf{X}}) + \mathrm{R}^{\ell}_{\mathcal{T}}(h^*_{\mathcal{S}}, h^*_{\mathcal{T}}) + \mathrm{R}^{\ell}_{\mathcal{T}}(h^*_{\mathcal{T}}, f_{\mathcal{T}}).\end{aligned}$$

□

THEOREM 3.5 ([COR 11, COR 14]).– *Let $\mathcal{S}$ and $\mathcal{T}$ be the source and the target domain on $\mathbf{X} \times Y$, $\mathcal{H}$ be a hypothesis class and ℓ be a μ-admissible loss. We assume that the target labeling function $f_{\mathcal{T}}$ belongs to $\mathcal{H}$, and let η denote $\max\{\ell(f_{\mathcal{S}}(\mathbf{x}), f_{\mathcal{T}}(\mathbf{x})) : \mathbf{x} \in \mathrm{SUPP}(\hat{\mathcal{S}}_{\mathbf{X}})\}$. Let h' be the hypothesis minimizing $F_{\hat{\mathcal{T}}_{\mathbf{X}}}$ and h the one returned when minimizing $F_{\hat{\mathcal{S}}_{\mathbf{X}}}$. Then, for all $(\mathbf{x}, y) \in \mathbf{X} \times Y$, we have*

$$|\ell(h'(\mathbf{x}), y) - \ell(h(\mathbf{x}), y)| \leq \mu R \sqrt{\frac{disc_{\ell}(\hat{\mathcal{S}}_{\mathbf{X}}, \hat{\mathcal{T}}_{\mathbf{X}}) + \mu\eta}{\beta}}.$$

PROOF.– In the proof, the authors make use of a generalized Bregman divergence that we introduce first. For a convex function $F : \mathbb{H} \to \mathbb{R}$, we denote by $\partial F(h)$ the subgradient of F at h: $\partial F(h) = \{g \in \mathcal{H} : \forall h' \in \mathbb{H}, F(h') - F(h) \geq \langle h' - h, g \rangle\}$ and note that $\partial F(h) = \nabla F(h)$ when F is differentiable at h and that for a point h where F is minimized, 0 is contained in $\partial F(h)$. As the subgradient is additive, i.e., for two convex functions, F_1 and F_2, $\partial F_1 + \partial F_2)(h) = \{g1 + g2 : g1 \in \partial F_1(h), g2 \in \partial F_2(h)\}$, then for any $h \in \mathbb{H}$, one may assume $\delta F(h)$ to be fixed to be an (arbitrary) element of $\partial F(h)$. For any such choice of δF, the generalized Bregman divergence associated with F can be defined by:

$$\forall\, h', h \in \mathbb{H},\ B_F(h' \| h) = F(h') - F(h) - \langle h' - h, \delta F(h) \rangle.$$

By definition of the subgradient, $B_F(h' \| h) \geq 0$ for all $h', h \in \mathbb{H}$. Let N denote the convex function $h \to \|h\|_K^2$. The convexity of N implies that it is differentiable, $\delta N(h) = \nabla N(h)$, $\forall h \in \mathbb{H}$, and δN and thus B_N are uniquely defined due to the property of a subgradient of a convex function. In order to define the Bregman divergences for $F_{\hat{\mathcal{S}}_{\mathbf{X}}}$ and $\mathrm{R}^{\ell}_{\hat{\mathcal{S}}_{\mathbf{X}}}$ so that $B_{F_{\hat{\mathcal{S}}_{\mathbf{X}}}} = B_{\mathrm{R}^{\ell}_{\hat{\mathcal{S}}_{\mathbf{X}}}} + \lambda B_N$, one can define $\delta \mathrm{R}^{\ell}_{\hat{\mathcal{S}}_{\mathbf{X}}}$ from $\delta B_{F_{\hat{\mathcal{S}}_{\mathbf{X}}}}$ by: $\delta \mathrm{R}^{\ell}_{\hat{\mathcal{S}}_{\mathbf{X}}}(h) = \delta F_{\hat{\mathcal{S}}_{\mathbf{X}}}(h) - \lambda \nabla N(h)$, $\forall h \in \mathbb{H}$. Furthermore, one can choose $\delta F_{\hat{\mathcal{S}}_{\mathbf{X}}}(h)$ to be 0 for any point h, where $F_{\hat{\mathcal{S}}_{\mathbf{X}}}(h)$ is minimized and let $\delta F_{\hat{\mathcal{S}}_{\mathbf{X}}}(h)$ be an arbitrary element of $\partial F_{\hat{\mathcal{S}}_{\mathbf{X}}}(h)$ for all other hypotheses. The Bregman divergences for $F_{\hat{\mathcal{T}}_{\mathbf{X}}}$ and $\mathrm{R}^{\ell}_{\hat{\mathcal{T}}_{\mathbf{X}}}$ can be defined in the same way so that $B_{F_{\hat{\mathcal{T}}_{\mathbf{X}}}} = B_{\mathrm{R}^{\ell}_{\hat{\mathcal{T}}_{\mathbf{X}}}} + \lambda B_N$.

Since the generalized Bregman divergence is non-negative and bearing in mind that $B_{F_{\hat{\mathcal{T}}_{\mathbf{X}}}} = B_{\mathrm{R}^{\ell}_{\hat{\mathcal{T}}_{\mathbf{X}}}} + \lambda B_N$ and $B_{F_{\hat{\mathcal{S}}_{\mathbf{X}}}} = B_{\mathrm{R}^{\ell}_{\hat{\mathcal{S}}_{\mathbf{X}}}} + \lambda B_N$, we obtain:

$$B_{F_{\hat{\mathcal{S}}_{\mathbf{X}}}}(h' \| h) + B_{F_{\hat{\mathcal{T}}_{\mathbf{X}}}}(h \| h') \geq \lambda (B_N(h' \| h) + B_N(h \| h')).$$

Note that $B_N(h' \| h) + B_N(h \| h') = -\langle h' - h, 2h \rangle - \langle h - h', 2h' \rangle = 2\|h' - h\|_K^2$. Thus, $B_N(h' \| h) + B_N(h \| h') \geq 2\|h' - h\|_K^2$. By definition of h' and h as minimizers and the choice of the subgradients, $\delta F_{\hat{\mathcal{T}}_{\mathbf{X}}}(h') = 0$ and $\delta F_{\hat{\mathcal{S}}_{\mathbf{X}}}(h) = 0$, this inequality can be rewritten as follows:

$$\begin{aligned}
2\lambda \|h' - h\|_K^2 &\leq \mathrm{R}^{\ell}_{\hat{\mathcal{S}}_{\mathbf{X}}}(h') - \mathrm{R}^{\ell}_{\hat{\mathcal{S}}_{\mathbf{X}}}(h) + \mathrm{R}^{\ell}_{\hat{\mathcal{T}}_{\mathbf{X}}}(h) - \mathrm{R}^{\ell}_{\hat{\mathcal{T}}_{\mathbf{X}}}(h') \\
&= (\mathrm{R}^{\ell}_{\hat{\mathcal{T}}_{\mathbf{X}}}(h, f_{\mathcal{T}}) - \mathrm{R}^{\ell}_{\hat{\mathcal{S}}_{\mathbf{X}}}(h, f_{\mathcal{S}})) - (\mathrm{R}^{\ell}_{\hat{\mathcal{T}}_{\mathbf{X}}}(h', f_{\mathcal{T}}) - \mathrm{R}^{\ell}_{\hat{\mathcal{S}}_{\mathbf{X}}}(h', f_{\mathcal{S}})) \\
&= (\mathrm{R}^{\ell}_{\hat{\mathcal{T}}_{\mathbf{X}}}(h, f_{\mathcal{T}}) - \mathrm{R}^{\ell}_{\hat{\mathcal{S}}_{\mathbf{X}}}(h, f_{\mathcal{T}})) - (\mathrm{R}^{\ell}_{\hat{\mathcal{T}}_{\mathbf{X}}}(h', f_{\mathcal{T}}) - \mathrm{R}^{\ell}_{\hat{\mathcal{S}}_{\mathbf{X}}}(h', f_{\mathcal{T}})) \\
&+ (\mathrm{R}^{\ell}_{\hat{\mathcal{S}}_{\mathbf{X}}}(h, f_{\mathcal{T}}) - \mathrm{R}^{\ell}_{\hat{\mathcal{S}}_{\mathbf{X}}}(h, f_{\mathcal{S}})) - (\mathrm{R}^{\ell}_{\hat{\mathcal{S}}_{\mathbf{X}}}(h', f_{\mathcal{T}}) - \mathrm{R}^{\ell}_{\hat{\mathcal{S}}_{\mathbf{X}}}(h', f_{\mathcal{S}}))
\end{aligned}$$

For $f_\mathcal{T} \in \mathcal{H}$, the first two terms can both be bounded by the empirical discrepancy:

$$|\mathrm{R}^\ell_{\hat{\mathcal{T}}_\mathbf{X}}(h, f_\mathcal{T}) - \mathrm{R}^\ell_{\hat{\mathcal{S}}_\mathbf{X}}(h, f_\mathcal{T})| \leq disc_\ell(\hat{\mathcal{S}}_\mathbf{X}, \hat{\mathcal{T}}_\mathbf{X}), |\mathrm{R}^\ell_{\hat{\mathcal{T}}_\mathbf{X}}(h', f_\mathcal{T}) - \mathrm{R}^\ell_{\hat{\mathcal{S}}_\mathbf{X}}(h', f_\mathcal{T})| \leq disc_\ell(\hat{\mathcal{S}}_\mathbf{X}, \hat{\mathcal{T}}_\mathbf{X}).$$

Using the μ-admissibility property of ℓ, the last two terms can be bounded for any $h'' \in \mathbb{H}$ as follows:

$$|\mathrm{R}^\ell_{\hat{\mathcal{S}}_\mathbf{X}}(h'', f_\mathcal{T}) - \mathrm{R}^\ell_{\hat{\mathcal{S}}_\mathbf{X}}(h'', f_\mathcal{S})| \leq \mu \mathop{\mathbf{E}}_{\mathbf{x}\sim\hat{\mathcal{S}}_\mathbf{X}} [|f_\mathcal{T}(\mathbf{x}) - f_\mathcal{S}(\mathbf{x})|] \leq \mu\eta.$$

Thus,

$$2\lambda\|h' - h\|_K^2 \leq 2disc_\ell(\hat{\mathcal{S}}_\mathbf{X}, \hat{\mathcal{T}}_\mathbf{X}) + 2\mu\eta. \qquad \text{[A1.13]}$$

By the reproducing property, for any $\mathbf{x} \in \mathbf{X}$, $h' - h(x) = \langle h'(x) - h, K(\mathbf{x}, \cdot)\rangle$, thus, for any $\mathbf{x} \in \mathbf{X}$ and $y \in Y$, $\ell(h'(x), y) - \ell(h(x), y) \leq \mu|h' - h| \leq \mu R\|h' - h\|_K$. Upper bounding the right-hand side using (A1.13) directly yields the final statement of the theorem. □

THEOREM 3.6 ([COR 14]).– *Let ℓ be a squared loss bounded by some $M > 0$ and let h' be the hypothesis minimizing $F_{\hat{\mathcal{T}}_\mathbf{X}}$ and h the one returned when minimizing $F_{\hat{\mathcal{S}}_\mathbf{X}}$. Then, for all $(\mathbf{x}, y) \in \mathbf{X} \times Y$, we have:*

$$|\ell(h(\mathbf{x}), y) - \ell(h'(\mathbf{x}), y)|$$

$$\leq \frac{R\sqrt{M}}{\beta}\left(\delta_\mathcal{H}(f_\mathcal{S}, f_\mathcal{T}) + \sqrt{\delta^2_\mathcal{H}(f_\mathcal{S}, f_\mathcal{T}) + 4\,\beta\, disc_\ell(\hat{\mathcal{S}}_\mathbf{X}, \hat{\mathcal{T}}_\mathbf{X})}\right).$$

PROOF.– Similar to the previous theorem, for any $h_0 \in \mathcal{H}$ we can write:

$$\begin{aligned} 2\lambda\|h' - h\|_K^2 \leq\ & \mathrm{R}^\ell_{\hat{\mathcal{T}}_\mathbf{X}}(h, f_\mathcal{T}) - \mathrm{R}^\ell_{\hat{\mathcal{T}}_\mathbf{X}}(h, h_0) - \mathrm{R}^\ell_{\hat{\mathcal{T}}_\mathbf{X}}(h', f_\mathcal{T}) - \mathrm{R}^\ell_{\hat{\mathcal{T}}_\mathbf{X}}(h', h_0) \\ & + \mathrm{R}^\ell_{\hat{\mathcal{T}}_\mathbf{X}}(h, h_0) - \mathrm{R}^\ell_{\hat{\mathcal{S}}_\mathbf{X}}(h, h_0) - \mathrm{R}^\ell_{\hat{\mathcal{T}}_\mathbf{X}}(h', f_\mathcal{T}) - \mathrm{R}^\ell_{\hat{\mathcal{S}}_\mathbf{X}}(h', h_0) \\ & + \mathrm{R}^\ell_{\hat{\mathcal{S}}_\mathbf{X}}(h, h_0) - \mathrm{R}^\ell_{\hat{\mathcal{S}}_\mathbf{X}}(h, f_\mathcal{S}) - \mathrm{R}^\ell_{\hat{\mathcal{S}}_\mathbf{X}}(h', h_0) - \mathrm{R}^\ell_{\hat{\mathcal{S}}_\mathbf{X}}(h', h_0). \end{aligned}$$

Then, the definition of the squared loss leads the following equalities:

$$\mathrm{R}^\ell_{\hat{\mathcal{T}}_\mathbf{X}}(h, f_\mathcal{T}) - \mathrm{R}^\ell_{\hat{\mathcal{T}}_\mathbf{X}}(h, h_0) = \mathop{\mathbf{E}}_{\mathbf{x}\sim\hat{\mathcal{T}}_\mathbf{X}} [(h_0(x) - f_\mathcal{T}(x))(2h(x) - f_\mathcal{T}(x) - h_0(x))]$$

$$\mathrm{R}^\ell_{\hat{\mathcal{T}}_\mathbf{X}}(h', f_\mathcal{T}) - \mathrm{R}^\ell_{\hat{\mathcal{T}}_\mathbf{X}}(h', h_0) = \mathop{\mathbf{E}}_{\mathbf{x}\sim\hat{\mathcal{T}}_\mathbf{X}} [(h_0(x) - f_\mathcal{T}(x))(2h'(x) - f_\mathcal{T}(x) - h_0(x))].$$

The difference between the left-hand sides of the last two lines gives:

$$(\mathrm{R}^{\ell}_{\hat{\mathcal{T}}_{\mathbf{X}}}(h, f_{\mathcal{T}}) - \mathrm{R}^{\ell}_{\hat{\mathcal{T}}_{\mathbf{X}}}(h, h_{\mathrm{O}})) - (\mathrm{R}^{\ell}_{\hat{\mathcal{T}}_{\mathbf{X}}}(h', f_{\mathcal{T}}) - \mathrm{R}^{\ell}_{\hat{\mathcal{T}}_{\mathbf{X}}}(h', h_{\mathrm{O}}))$$
$$= 2 \mathop{\mathbf{E}}_{\mathbf{x} \sim \hat{\mathcal{T}}_{\mathbf{X}}} [(h_{\mathrm{O}}(x) - f_{\mathcal{T}}(x))(h(x) - h'(x))].$$

Similarly, for the source domain we obtain:

$$(\mathrm{R}^{\ell}_{\hat{\mathcal{S}}_{\mathbf{X}}}(h, h_{\mathrm{O}}) - \mathrm{R}^{\ell}_{\hat{\mathcal{S}}_{\mathbf{X}}}(h, f_{\mathcal{S}})) - (\mathrm{R}^{\ell}_{\hat{\mathcal{S}}_{\mathbf{X}}}(h', h_{\mathrm{O}}) - \mathrm{R}^{\ell}_{\hat{\mathcal{S}}_{\mathbf{X}}}(h', f_{\mathcal{S}}))$$
$$= -2 \mathop{\mathbf{E}}_{\mathbf{x} \sim \hat{\mathcal{S}}_{\mathbf{X}}} [(h_{\mathrm{O}}(x) - f_{\mathcal{S}}(x))(h(x) - h'(x))].$$

Using the definition of the discrepancy, we bound the following difference:

$$(\mathrm{R}^{\ell}_{\hat{\mathcal{T}}_{\mathbf{X}}}(h, h_{\mathrm{O}}) - \mathrm{R}^{\ell}_{\hat{\mathcal{S}}_{\mathbf{X}}}(h, h_{\mathrm{O}})) - (\mathrm{R}^{\ell}_{\hat{\mathcal{T}}_{\mathbf{X}}}(h', h_{\mathrm{O}}) - \mathrm{R}^{\ell}_{\hat{\mathcal{S}}_{\mathbf{X}}}(h', h_{\mathrm{O}})) \leq 2disc_{\ell}(\hat{\mathcal{S}}_{\mathbf{X}}, \hat{\mathcal{T}}_{\mathbf{X}}).$$

Applying this inequality to the norm of the difference between hypotheses h and h' gives:

$$2\lambda \|h' - h\|^2_K \leq 2disc_{\ell}(\hat{\mathcal{S}}_{\mathbf{X}}, \hat{\mathcal{T}}_{\mathbf{X}}) + 2\Delta,$$

where

$$\Delta = \mathop{\mathbf{E}}_{\mathbf{x} \sim \hat{\mathcal{T}}_{\mathbf{X}}} [(h_{\mathrm{O}}(x) - f_{\mathcal{T}}(x)(h(x) - h'(x))] - \mathop{\mathbf{E}}_{\mathbf{x} \sim \hat{\mathcal{S}}_{\mathbf{X}}} [(h_{\mathrm{O}}(x) - f_{\mathcal{S}}(x)(h(x) - h'(x))].$$

We can now use the reproducing property applied to the difference $h - h'$ to obtain the identity $h - h' = \langle h - h', K(\mathbf{x}, \cdot)\rangle$, $\forall \mathbf{x} \in \mathbf{X}$. We now rewrite

$$\Delta = \langle h - h', \mathop{\mathbf{E}}_{\mathbf{x} \sim \hat{\mathcal{T}}_{\mathbf{X}}} [(h_{\mathrm{O}}(x) - f_{\mathcal{T}}(x))K(\mathbf{x}, \cdot)] - \mathop{\mathbf{E}}_{\mathbf{x} \sim \hat{\mathcal{S}}_{\mathbf{X}}} [(h_{\mathrm{O}}(x) - f_{\mathcal{S}}(x))K(\mathbf{x}, \cdot)]\rangle$$
$$\leq \|h' - h\| \| \mathop{\mathbf{E}}_{\mathbf{x} \sim \hat{\mathcal{T}}_{\mathbf{X}}} [(h_{\mathrm{O}}(x) - f_{\mathcal{T}}(x))K(\mathbf{x}, \cdot)] - \mathop{\mathbf{E}}_{\mathbf{x} \sim \hat{\mathcal{S}}_{\mathbf{X}}} [(h_{\mathrm{O}}(x) - f_{\mathcal{S}}(x))K(\mathbf{x}, \cdot)]\|$$
$$\leq \|h' - h\| \delta_{\mathcal{H}}(f_{\mathcal{S}}, f_{\mathcal{T}}).$$

The last result in its turn, leads to the following inequality:

$$\|h' - h\| \leq \frac{1}{2\lambda} \left(\delta_{\mathcal{H}}(f_{\mathcal{S}}, f_{\mathcal{T}}) + \sqrt{\delta^2_{\mathcal{H}}(f_{\mathcal{S}}, f_{\mathcal{T}}) + 4\lambda disc_{\ell}(\hat{\mathcal{S}}_{\mathbf{X}}, \hat{\mathcal{T}}_{\mathbf{X}})} \right).$$

The final result is obtained by using the property of the squared loss:

$$|\ell(h(\mathbf{x}), y) - \ell(h'(\mathbf{x}), y)| = |(h(\mathbf{x}) - y)^2 - (h'(\mathbf{x}) - y)^2|$$
$$= |(h(\mathbf{x}) - h'(\mathbf{x}))(h'(x) - y + h(x) - y)|$$

$$\leq 2\sqrt{M}|h'(x) - h(x)|$$
$$= 2\sqrt{M}|\langle h - h', K(\mathbf{x}, \cdot)\rangle| \leq 2\sqrt{M}R\|h' - h\|.$$

□

THEOREM 3.7 ([COR 14]).– *Assume that for all* $\mathbf{x} \in \mathbf{X}$, $K(\mathbf{x}, \mathbf{x}) \leq R^2$ *for some* $R > 0$. *Let* $\mathcal{A}$ *denote the union of the supports of* $\hat{\mathcal{S}}_\mathbf{X}$ *and* $\hat{\mathcal{T}}_\mathbf{X}$. *Then, for any* $p > 1$ *and* $q > 1$, *with* $1/p + 1/q = 1$,

$$\delta_\mathcal{H}(f, f) \leq d_p(f_{|\mathcal{A}}, \mathcal{H}_{|\mathcal{A}})\ell_q(\hat{\mathcal{S}}_\mathbf{X}, \hat{\mathcal{T}}_\mathbf{X}),$$

where for any set $\mathcal{A} \subseteq \mathbf{X}$, $f_{|\mathcal{A}}$ *(respectively,* $\mathcal{H}_{|\mathcal{A}}$*) denote the restriction of* f *(respectively,* h*) to* $\mathcal{A}$ *and* $d_p(f_{|\mathcal{A}}, \mathcal{H}_{|\mathcal{A}}) = \inf_{h\in\mathcal{H}} \|f - h\|_p$.

PROOF.– We can write the left-hand side as follows:

$$\delta_\mathcal{H}(f) = \inf_{h\in\mathcal{H}} \|\sum_{\mathbf{x}\in\mathcal{A}} (\hat{\mathcal{T}}_\mathbf{X}(\mathbf{x}) - \hat{\mathcal{S}}_\mathbf{X}(\mathbf{x}))(h(\mathbf{x}) - f(\mathbf{x}))\Phi(\mathbf{x})\|_K$$
$$\leq \inf_{h\in\mathcal{H}} \sum_{\mathbf{x}\in\mathcal{A}} |\hat{\mathcal{T}}_\mathbf{X}(\mathbf{x}) - \hat{\mathcal{S}}_\mathbf{X}(\mathbf{x})||h(\mathbf{x}) - f(\mathbf{x})|\|\Phi(\mathbf{x})\|_K$$
$$\leq R \inf_{h\in\mathcal{H}} \sum_{\mathbf{x}\in\mathcal{A}} |\hat{\mathcal{T}}_\mathbf{X}(\mathbf{x}) - \hat{\mathcal{S}}_\mathbf{X}(\mathbf{x})|\, |h(\mathbf{x}) - f(\mathbf{x})|\mathbf{x}.$$

The final result is obtained by applying Hölder's inequality:

$$\delta_\mathcal{H}(f) = R\ell_q(\hat{\mathcal{S}}_\mathbf{X}, \hat{\mathcal{T}}_\mathbf{X})\|f_{|\mathcal{A}}, \mathcal{H}_{|\mathcal{A}}\|_p \leq d_p(f_{|\mathcal{A}}, \mathcal{H}_{|\mathcal{A}})\ell_q(\hat{\mathcal{S}}_\mathbf{X}, \hat{\mathcal{T}}_\mathbf{X}).$$

□

THEOREM 3.8 ([COR 15]).– *Let* h^* *be a minimizer of* $\mathrm{R}^\ell_{\hat{\mathcal{T}}_\mathbf{X}}(h, f_\mathcal{T}) + \beta\|h\|^2_K$ *and* h_Q *be a minimizer of* $\mathrm{R}^\ell_{Q_h}(h, f_\mathcal{S}) + \beta\|h\|^2_K$. *Then, for* $Q : h \to Q_h$ *and* $\forall\mathbf{x} \in \mathbf{X}$, $y \in Y$, *the following holds:*

$$|\ell(h_Q(\mathbf{x}), y) - \ell(h^*(\mathbf{x}), y)| \leq \mu R\sqrt{\frac{\mu d^{\hat{\mathcal{T}}_\mathbf{X}}_\infty(f_\mathcal{T}, \mathcal{H}'') + DISC(Q, \hat{\mathcal{T}}_\mathbf{X})}{\beta}}.$$

PROOF.– In order to prove this result, the authors first establish a bound on the difference between the optimal hypotheses in terms of the generalized discrepancy. It can be stated as follows:

LEMMA A1.1.– *Let* U *be an arbitrary element in* $A(\mathcal{H})$ *and* h^* *and* h_U *be the hypotheses minimizing* $\mathrm{R}^\ell_{\hat{\mathcal{T}}_\mathbf{X}}(h, f_\mathcal{T}) + \beta\|h\|^2_K$ *and* $\mathrm{R}^\ell_{U_h}(h, f_\mathcal{S}) + \beta\|h\|^2_K$, *respectively. Then, the following inequality holds for any convex set* $\mathcal{H}'' \subseteq \mathcal{H}$:

$$\lambda\|h^* - h_U\|^2_K \leq \mu d^{\hat{\mathcal{T}}_\mathbf{X}}_\infty(f_\mathcal{T}, \mathcal{H}'') + DISC(Q, \hat{\mathcal{T}}_\mathbf{X}).$$

To proceed, we fix an element U from $A(\mathcal{H})$ and let $G_{\hat{\mathcal{T}}_\mathbf{X}}$ denote the mapping $h \to \mathrm{R}^{\ell}_{\hat{\mathcal{T}}_\mathbf{X}}(h, f_{\mathcal{T}})$. We further define G_U to be the function $h \to \mathrm{R}^{\ell}_{U_h}(h, f_{\mathcal{S}})$. As $h \to \lambda\|h\|^2_K + G_{\hat{\mathcal{T}}_\mathbf{X}}(h)$ is convex and differentiable and since h^* is a minimizer of $G_{\hat{\mathcal{T}}_\mathbf{X}}$, the gradient is zero at h^* meaning that $2\lambda h^* = -\nabla G_{\hat{\mathcal{T}}_\mathbf{X}}(h^*)$. However, since $h \to \lambda\|h\|^2_K + G_U(h)$ is convex, it admits a subdifferential at any $h \in \mathcal{H}$. As h_U denotes a minimizer, its subdifferential at h_U inevitably contains 0. Thus, there exists a subgradient $g_0 \in \partial G_U(h_U)$ such that $2\lambda h_U = -g_0$, where $\partial G_U(h_U)$ denotes the subdifferential of G_U at h_U. Using these two equalities, we obtain:

$$
\begin{aligned}
2\lambda\|h^* - h_U\|^2_K &= \langle h^* - h_U, g_0 - \nabla G_{\hat{\mathcal{T}}_\mathbf{X}}(h^*)\rangle \\
&= \langle g_0, h^* - h_U\rangle - \langle \nabla G_{\hat{\mathcal{T}}_\mathbf{X}}(h^*), h^* - h_U\rangle \\
&\leq G_U(h^*) - G_U(h_U) + G_{\hat{\mathcal{T}}_\mathbf{X}}(h_U) - G_{\hat{\mathcal{T}}_\mathbf{X}}(h^*) \\
&= \mathrm{R}^{\ell}_{\hat{\mathcal{T}}_\mathbf{X}}(h_U, f_{\mathcal{T}}) - \mathrm{R}^{\ell}_{U_h}(h_U, f_{\mathcal{S}}) + \mathrm{R}^{\ell}_{U_h}(h^*, f_{\mathcal{S}}) - \mathrm{R}^{\ell}_{\hat{\mathcal{T}}_\mathbf{X}}(h^*, f_{\mathcal{T}}) \\
&\leq 2 \max_{h\in\mathcal{H}} \left|\mathrm{R}^{\ell}_{\hat{\mathcal{T}}_\mathbf{X}}(h, f_{\mathcal{T}}) - \mathrm{R}^{\ell}_{U_h}(h, f_{\mathcal{S}})\right|,
\end{aligned}
$$

where the first inequality is due to the convexity of G_U in combination with the subgradient property of $g_0 \in \partial G_U(h_U)$, and the convexity of $G_{\hat{\mathcal{T}}_\mathbf{X}}$. Using the μ-admissibility of the loss, we can upper bound the difference $\mathrm{R}^{\ell}_{\hat{\mathcal{T}}_\mathbf{X}}(h, f_{\mathcal{T}}) - \mathrm{R}^{\ell}_{U_h}(h, f_{\mathcal{S}})$ for any $h \in \mathcal{H}$ as follows:

$$
\begin{aligned}
\|\mathrm{R}^{\ell}_{\hat{\mathcal{T}}_\mathbf{X}}(h, f_{\mathcal{T}}) - \mathrm{R}^{\ell}_{U_h}(h, f_{\mathcal{S}})\| &\leq \left|\mathrm{R}^{\ell}_{\hat{\mathcal{T}}_\mathbf{X}}(h, f_{\mathcal{T}}) - \mathrm{R}^{\ell}_{\hat{\mathcal{T}}_\mathbf{X}}(h, h_0)\right| \\
&\quad + \left|\mathrm{R}^{\ell}_{\hat{\mathcal{T}}_\mathbf{X}}(h, h_0) - \mathrm{R}^{\ell}_{U_h}(h, f_{\mathcal{S}})\right| \\
&\leq \mu \mathop{\mathbf{E}}_{\mathbf{x}\sim\hat{\mathcal{T}}_\mathbf{X}} |f_{\mathcal{T}}(x) - h_0(x)| + \max_{h''\in\mathcal{H}''} \left|\mathrm{R}^{\ell}_{\hat{\mathcal{T}}_\mathbf{X}}(h, h'') - \mathrm{R}^{\ell}_{U_h}(h, f_{\mathcal{S}})\right| \\
&\leq \mu \max_{\mathbf{x}\in\mathrm{SUPP}(\hat{\mathcal{T}}_\mathbf{X})} |f_{\mathcal{T}}(x) - h_0(x)| + \max_{h''\in\mathcal{H}''} \left|\mathrm{R}^{\ell}_{\hat{\mathcal{T}}_\mathbf{X}}(h, h'') - \mathrm{R}^{\ell}_{U_h}(h, f_{\mathcal{S}})\right|,
\end{aligned}
$$

where h_0 is an arbitrary element of $\mathcal{H}''$. As this bound holds for all $h_0 \in \mathcal{H}''$, we can obtain the final result as:

$$
\begin{aligned}
&\lambda\|h^* - h_U\|^2_K \\
&\leq \mu \min_{h_0\in\mathcal{H}''} \max_{\mathbf{x}\in\mathrm{SUPP}(\hat{\mathcal{T}}_\mathbf{X})} |f_{\mathcal{T}}(x) - h_0(x)| + \max_{h\in\mathcal{H}} \max_{h''\in\mathcal{H}''} \left|\mathrm{R}^{\ell}_{\hat{\mathcal{T}}_\mathbf{X}}(h, h'') - \mathrm{R}^{\ell}_{U_h}(h, f_{\mathcal{S}})\right|.
\end{aligned}
$$

This proves the lemma. Now, by the μ-admissibility of the loss, using the reproducing property of $\mathcal{H}$, and the Cauchy–Schwarz inequality, the following holds for all $\mathbf{x} \in \mathbf{X}$ and $y \in Y$:

$$|\ell(h_Q(\mathbf{x}), y) - \ell(h^*(\mathbf{x}), y)| \leq \mu|h' - h| \leq |\langle h' - h, K(\mathbf{x}, \cdot)\rangle|$$
$$\|h' - h^*\|_K \sqrt{K(\mathbf{x}, \mathbf{x})} \leq R\|h' - h^*\|.$$

The desired result follows from upper bounding $\|h' - h^*\|_K$ using the bound of the theorem established above and using the fact that Q is a minimizer of the bound over all choices of $U \in A(\mathcal{H})$. □

Appendix 2

Proofs of the Main Results of Chapter 4

THEOREM 4.1 (Necessity of small $\mathcal{H}\Delta\mathcal{H}$-distance [BEN 10b]).– *Let* $\mathbf{X}$ *be some domain set and* $\mathcal{H}$ *a class of functions over* $\mathbf{X}$*. Assume that, for some* $\mathcal{A} \subseteq \mathbf{X}$*, we have that* $\{h^{-1}(1) \cap \mathcal{A} : h \in \mathcal{H}\}$ *contains more than two sets and is linearly ordered by inclusion. Then, the conditions covariate shift plus small* $\lambda_{\mathcal{H}}$ *do not suffice for domain adaptation. In particular, for every* $\epsilon > 0$ *there exists probability distributions* $\mathcal{S}$ *over* $\mathbf{X} \times \{0,1\}$*, and* $\mathcal{T}_{\mathbf{X}}$ *over* $\mathbf{X}$ *such that for every domain adaptation learner* $\mathcal{A}$*, all integers* $m > 0$*,* $n > 0$*, there exists a labeling function* $f : \mathbf{X} \to \{0,1\}$ *such that*

1) $\lambda_{\mathcal{H}} \leq \epsilon$ *is small;*

2) $\mathcal{S}$ *and* $\mathcal{T}_f$ *satisfy the covariate shift assumption;*

3) $\underset{\substack{S \sim (\mathcal{S})^m \\ T_u \sim (\mathcal{T}_{\mathbf{X}})^n}}{\mathbf{Pr}} \left[\mathrm{R}_{\mathcal{T}_f}(\mathcal{A}(S, T_u)) \geq \frac{1}{2} \right] \geq \frac{1}{2}$,

where the distribution $\mathcal{T}_f$ *over* $\mathbf{X} \times \{0,1\}$ *is defined as* $\mathcal{T}_f\{1|\mathbf{x} \in \mathbf{X}\} = f(\mathbf{x})$.

PROOF.– The idea of this proof is to find an example of source and target distributions for which the same input of a DA learner can lead to completely different results in terms of the error. To this end, the authors construct a labeled distribution $\mathcal{S}$ and an unlabeled distribution $\mathcal{T}$ over $\mathbf{X}$, as well as two labeling functions g and g' so that the triples $(\mathcal{S}, \mathcal{T}_g, \mathcal{H})$ and $(\mathcal{S}, \mathcal{T}_{g'}, \mathcal{H})$ are indistinguishable, even though any hypothesis that has small error on $\mathcal{T}_g$ leads to a very poor performance on $\mathcal{T}_{g'}$ and vice versa. The idea behind this is to make sure that the learner $\mathcal{A}$ is not able to tell which of the two target distributions it is trying to learn, meaning that any of its output hypotheses have an error of at least 0.5 on one of the possible target distributions. Then, for any $\mathcal{A}$, one can pick the target on which it fails. This proof considers the setting for the case, where the input space $\mathbf{X}$ is assumed to be defined on the real line, while $\mathcal{H}$ consists solely of the threshold function. Obviously, the obtained results can easily be extended to the case of a general $\mathcal{H}$ stated in the theorem.

To proceed, we consider that $\mathcal{S}_{\mathbf{X}}$ is uniform in the interval $[0,1]$ and takes the values in the interval $[1-\epsilon, 1]$ given by "odd points" $\{(2k+1)\xi : k \in \mathbb{N}, 1-\epsilon \leq (2k+1)\xi \leq 1\}$. Similarly, for $\mathcal{T}$, we assume that it is uniform over $[0,1]$ and takes the values in the interval $[1-\epsilon, 1]$ given by "even points" $\{2k\xi : k \in \mathbb{N}, 1-\epsilon \leq 2k\xi \leq 1\}$. The source distribution $\mathcal{S}$ is labeled with 0's everywhere except in the interval $[1-\epsilon, 1]$, where the "odd points" are labeled as 1 (the rest of the points in that interval will be labeled by either g or g'). In this case, the covariate shift assumption (2) is satisfied by observing that the supports of $\mathcal{S}$ and $\mathcal{T}$ are disjoint and by defining g, g' as in agreement with $\mathcal{S}$ on its support, while on the support of $\mathcal{T}$ (the "even points") g and g' attribute labels 1 and 0, respectively. This last construction makes $\lambda_{\mathcal{H}}$ small regardless of whether g or g' is used as the true labeling function f, thus satisfying the first condition of the theorem. One may note that in this case for any two samples L, U, the actual output of $\mathcal{A}$ leads to the sum of the errors (with respect to $\mathcal{T}_g$ and $\mathcal{T}_{g'}$) equal to 1. If we suppose that $\mathcal{A}$ is deterministic, then we can analyze the following two sets:

$$G = \left\{(L,U) : |L| = m, |U| = n, \mathrm{R}_{\mathcal{T}_g}(\mathcal{A}(L,U)) > \frac{1}{2}\right\},$$

$$G' = \left\{(L,U) : |L| = m, |U| = n, \mathrm{R}_{\mathcal{T}_{g'}}(\mathcal{A}(L,U)) > \frac{1}{2}\right\}.$$

Obviously, G and G' have disjoint union given by the set of all (L,U) with $|L| = m, |U| = n$. If one chooses $f = g$ in the case where $\mathbf{Pr}\,[G] \geq \frac{1}{2}$ and g' otherwise, then (3) follows immediately. The same reasoning can be applied to provide this result in case of broader hypothesis classes, such as linear, homogeneous half-spaces in $\mathbb{R}^d$. □

THEOREM 4.2 (Necessity of small $\lambda_{\mathcal{H}}$ [BEN 10b]).– *Let* $\mathbf{X}$ *be some domain set, and* $\mathcal{H}$ *a class of functions over* $\mathbf{X}$ *whose VC dimension is much smaller than* $|\mathbf{X}|$ *(for instance, any* $\mathcal{H}$ *with a finite VC dimension over an infinite* $\mathbf{X}$*). Then, the conditions covariate shift plus small* $\mathcal{H}\Delta\mathcal{H}$*-divergence do not suffice for domain adaptation. In particular, for every* $\epsilon > 0$ *there exists probability distributions* $\mathcal{S}$ *over* $\mathbf{X} \times \{0,1\}$*, and* $\mathcal{T}_{\mathbf{X}}$ *over* $\mathbf{X}$ *such that for every domain adaptation learner* $\mathcal{A}$*, all integers* $m, n > 0$*, there exists a labeling function* $f : \mathbf{X} \to \{0,1\}$ *such that*

1) $d_{\mathcal{H}\Delta\mathcal{H}}(\mathcal{T}_{\mathbf{X}}, \mathcal{S}_{\mathbf{X}}) \leq \epsilon$ *is small;*

2) the covariate shift assumption holds;

3) $\underset{\substack{S \sim \mathcal{S}^m \\ T_u \sim (\mathcal{T}_{\mathbf{X}})^n}}{\mathbf{Pr}} \left[\mathrm{R}_{\mathcal{T}_f}(\mathcal{A}(S, T_u)) \geq \frac{1}{2}\right] \geq \frac{1}{2}.$

PROOF.– The proof follows the same reasoning as that of the previous theorem. Consider $\mathcal{T}$ to be a uniform distribution over $\{2k\xi : k \in \mathbb{N}, 2k\xi \leq 1\} \times \{1\}$ and let the source distribution $\mathcal{S}$ be the uniform distribution over $\{(2k+1)\xi : k \in \mathbb{N}, (2k+1)\xi \leq 1\} \times \{0\}$, such that (1) is satisfied. Now, we can construct two target functions g, g' that both agree with $\mathcal{S}$ on its support, but on support of $\mathcal{T}$, g labels all 1's and g' labels 0. Note that covariate shift holds regardless of whether one chooses f to be equal to g or g'. Furthermore, the learner $\mathcal{A}$ is not able to tell whether g or g' is the true labeling function on $\mathcal{T}$ and makes a total error of 1 on $\mathcal{T}_g$ and $\mathcal{T}_{g'}$ combined. Using similar arguments to in the proof above, we can also establish (3). □

THEOREM 4.3 ([BEN 12c]).– *For every finite domain* $\mathbf{X}$, *for every* ε *and* δ *with* $\varepsilon + \delta < \frac{1}{2}$, *no algorithm can* $(\varepsilon, \delta, s, t)$*-solve the domain adaptation problem for the class* $\mathcal{W}$ *of triples* $(\mathcal{S}_\mathbf{X}, \mathcal{T}_\mathbf{X}, f)$ *with* $C_\mathcal{B}(\mathcal{S}_\mathbf{X}, \mathcal{T}_\mathbf{X}) \geq \frac{1}{2}$, $d_{\mathcal{H}\Delta\mathcal{H}}(\mathcal{S}_\mathbf{X}, \mathcal{T}_\mathbf{X}) = 0$ *and* $\mathrm{R}_\mathcal{T}(\mathcal{H}) = 0$ *if*

$$s + t < \sqrt{(1 - 2(\varepsilon + \delta))|\mathbf{X}|},$$

where $\mathcal{H}$ *is the hypothesis class that contains only the all-1 and the all-0 labeling functions,* $\mathrm{R}_\mathcal{T}(\mathcal{H}) = \min_{h \in \mathcal{H}} \mathrm{R}_\mathcal{T}(h, f)$ *and*

$$C_\mathcal{B}(\mathcal{S}_\mathbf{X}, \mathcal{T}_\mathbf{X}) = \inf_{\substack{b \in \mathcal{B} \\ \mathcal{T}_\mathbf{X}(b) \neq 0}} \frac{\mathcal{S}_\mathbf{X}(b)}{\mathcal{T}_\mathbf{X}(b)}$$

is a weight ratio [BEN 12c] of source and target domains, with respect to a collection of input space subsets $\mathcal{B} \subseteq 2^\mathbf{X}$.

PROOF.– The lower bound presented in this theorem was obtained by the authors as a result of reducing the left/right problem [BAT 13] to domain adaptation. In this problem, the task considered is to distinguish two distributions from finite samples where the input consists of three finite samples L, R and M of points from some domain set $\mathbf{X}$. Assuming that L is an i.i.d. sample from some distribution P over $\mathbf{X}$, R is an an i.i.d. sample from some distribution Q over $\mathbf{X}$ and M is an i.i.d. sample generated by one of these two probability distributions, was M generated by P or by Q?

To proceed, we first present a lower bound on the sample size needed to solve the left/right problem in the following lemma.

LEMMA A2.1.– *For any given sample sizes* l *for* L, r *for* R *and* m *for* M *and any* $0 < \gamma < \frac{1}{2}$, *if* $k = \max\{l, r\} + m$, *then for* $n > (k+1)^2/1 - 2\gamma$ *no algorithm gives the correct answer to the left/right problem with probability greater than* $1 - \gamma$ *over the class* $W_n^{uni} = \{(U_A, U_B, U_C) : A \cup B = \{1, \ldots, n\}, A \cap B = \emptyset, |A| = |B|,$ *and* $C = A$ *or* $C = B\}$, *where, for a finite set* Y, U_Y *denotes the uniform distribution over* Y.

In order to reduce the left/right problem to domain adaptation, the starting point is to define a class of DA problems that corresponds to the class of triples W_n^{uni}, for which a lower bound on the sample sizes needed for solving the left/right problem was obtained. To proceed, let W_n^{DA} be the class of triples $(\mathcal{S}_{\mathbf{X}}, \mathcal{T}_{\mathbf{X}}, l)$ for a number n, where $\mathcal{S}_{\mathbf{X}}$ is uniform over some finite set $\mathbf{X}$ of size n, $\mathcal{T}_{\mathbf{X}}$ is uniform over some subset U of $\mathbf{X}$ of size $n/2$ and f assigns points in U to 1 and points in $\mathbf{X} \setminus U$ to 0 or vice versa. From this construction, we may note that the weight ratio $C(\mathcal{S}_{\mathbf{X}}, \mathcal{T}_{\mathbf{X}})$ is equal to $\frac{1}{2}$ and $d_{\mathcal{H}\Delta\mathcal{H}}(\mathcal{S}_{\mathbf{X}}, \mathcal{T}_{\mathbf{X}}) = 0$ for all $(\mathcal{S}_{\mathbf{X}}, \mathcal{T}_{\mathbf{X}}, f)$ in W_n^{DA}. Furthermore, for the class $\mathcal{H}$ that consists of only the constant 1 and 0 functions, we have $\mathrm{R}^*_{\mathcal{T}}(\mathcal{H}) = 0$ for all elements of W_n^{DA}.

The two problems can be now related by the following result.

LEMMA A2.2.– *The left/right problem reduces to domain adaptation. More precisely, for given a number* n *and an algorithm* $\mathcal{A}$ *that, given the promise that the target task is realizable by the class* $\mathcal{H}$, *can* $(\varepsilon, \delta, s, t)$*-solve DA for a class* W *that includes* W_n^{DA}, *we can construct an algorithm that* $(\varepsilon + \delta, s, s, t + 1)$*-solves the left/right problem on* W_n^{uni}.

Given a triple (U_A, U_B, U_C) of distributions in W_n^{uni}, let us assume that we have access to samples $L = l_1, l_2, \cdots, l_s$ and $R = r_1, r_2, \cdots, r_s$ of size s and a sample M of size $t+1$ for the left/right problem coming from it. We can now obtain an input to the DA problem by choosing the unlabeled target sample $T = M \setminus \{p\}$, where p is a point from M drawn uniformly at random. We further construct the labeled source sample S by picking s elements from $L \times \{0\} \cup R \times \{1\}$ where the selection is obtained by successively flipping an unbiased coin and depending on the result choosing the next element from $L \times \{0\}$ or $R \times \{1\}$.

This construction can now be used as an input for a DA problem generated from a source distribution $\mathcal{S}_{\mathbf{X}} = U_{A \cup B}$ that is uniform over $A \cup B$. The target distribution $\mathcal{T}_{\mathbf{X}}$ in this case has the target marginal distribution equal to U_A or to U_B (depending on whether M was a sample from U_A or from U_B). Furthermore, the labeling function in this scenario is $f(\mathbf{x}) = 0$ if $\mathbf{x} \in A$ and $f(\mathbf{x}) = 1$ if $\mathbf{x} \in B$. As mentioned before, we have $C(\mathcal{S}_{\mathbf{X}}, \mathcal{T}_{\mathbf{X}}) = \frac{1}{2}$ and $d_{\mathcal{H}\Delta\mathcal{H}}(\mathcal{S}_{\mathbf{X}}, \mathcal{T}_{\mathbf{X}}) = 0$ for all $(\mathcal{S}_{\mathbf{X}}, \mathcal{T}_{\mathbf{X}}, f)$ in W_n^{DA}. Let us denote by h the output produced by $\mathcal{A}$ when taken at input S and T. The algorithm for the left/right problem then outputs U_A if $h(p) = 0$ and U_B if $h(p) = 1$ and the final results follow as $\mathrm{R}_{\mathcal{S}_{\mathbf{X}}}(h) \leq \epsilon$ with confidence $1 - \delta$. This proves the above lemma. The two lemmas thus show that no algorithm can solve the DA problem for W_n^{DA}, even if $\mathcal{H}$ contains a hypothesis that achieves zero error, if the sample sizes of the source and target sample satisfy $|S| + |T| < (1 - 2(\epsilon + \delta))\,|\mathbf{X}|$. This yields the claim of the main theorem. □

THEOREM 4.4 ([BEN 12c]).– *Let* $\mathbf{X} = [0,1]^d$, $\varepsilon > 0$ *and* $\delta > 0$ *be such that* $\varepsilon + \delta < \frac{1}{2}$, $\lambda > 1$ *and* $\mathcal{W}_\lambda$ *be the set of triples* $(\mathcal{S}_\mathbf{X}, \mathcal{T}_\mathbf{X}, f)$ *of distributions over* $\mathbf{X}$ *with* $\mathrm{R}_\mathcal{T}(\mathcal{H}) = 0$, $C_\mathcal{B}(\mathcal{S}_\mathbf{X}, \mathcal{T}_\mathbf{X}) \geq \frac{1}{2}$, $d_{\mathcal{H}\Delta\mathcal{H}}(\mathcal{S}_\mathbf{X}, \mathcal{T}_\mathbf{X}) = 0$ *and* λ*-Lipschitz labeling functions* f. *Then no domain adaptation-learner can* $(\varepsilon, \delta, s, t)$*-solve the domain adaptation problem for the class* $\mathcal{W}_\lambda$ *unless*

$$s + t \geq \sqrt{(\lambda+1)^d(1 - 2(\varepsilon+\delta))}.$$

PROOF.– We assume that G is a subset of $\mathbf{X}$ consisting of the points of a grid in $[0,1]^d$ with distance $1/\lambda$. For this particular setting, we have $|G| = (\lambda+1)^d$. From the statement of the theorem, the class $\mathcal{W}_\lambda$ contains all triples $(\mathcal{S}_\mathbf{X}, \mathcal{T}_\mathbf{X}, f)$, where the support of $\mathcal{S}_\mathbf{X}$ and $\mathcal{T}_\mathbf{X}$ is G, $\mathrm{R}^*_\mathcal{T}(\mathcal{H}) = 0$, $C_\mathcal{B}(\mathcal{S}_\mathbf{X}, \mathcal{T}_\mathbf{X}) \geq \frac{1}{2}$, $d_{\mathcal{H}\Delta\mathcal{H}}(\mathcal{S}_\mathbf{X}, \mathcal{T}_\mathbf{X}) = 0$ and arbitrary labeling functions $f : G \to \{0,1\}$, as every such function is λ-Lipschitz. As G is finite, the bound follows from theorem 4.3. □

THEOREM 4.5 ([BEN 12c]).– *Let* $\mathbf{X} = [0,1]^d$, $\gamma > 0$ *be a margin parameter,* $\mathcal{H}$ *be a hypothesis class of finite VC dimension and* $\mathcal{W}$ *be the set of triples* $(\mathcal{S}_\mathbf{X}, \mathcal{T}_\mathbf{X}, f)$ *of the source distribution, target distribution and labeling function with*

1) $C_\mathcal{I}(\mathcal{S}_\mathbf{X}, \mathcal{T}_\mathbf{X}) \geq \frac{1}{2}$ *for the class* $\mathcal{I} = (\mathcal{H}\Delta\mathcal{H}) \cap \mathcal{B}$, *where* $\mathcal{B}$ *is a partition of* $[0,1]^d$ *into boxes of sidelength* $\frac{\gamma}{\sqrt{d}}$;

2) $\mathcal{H}$ *contains a hypothesis that has* γ*-margin on* $\mathcal{T}$;

3) the labeling function f *is a* γ*-margin classifier with respect to* $\mathcal{T}$.

Then there is a constant $c > 1$ *such that, for all* $\varepsilon > 0$, $\delta > 0$ *and for all* $(\mathcal{S}_\mathbf{X}, \mathcal{T}_\mathbf{X}, f) \in \mathcal{W}$, *when given an i.i.d. sample* S_u *from* $\mathcal{S}_\mathbf{X}$, *labeled by* f *of size*

$$|S_u| \geq c\left[\frac{VC(\mathcal{H}) + \log\frac{1}{\delta}}{C_\mathcal{I}(\mathcal{S}_\mathbf{X}, \mathcal{T}_\mathbf{X})(1-\varepsilon)\varepsilon} \log\left(\frac{VC(\mathcal{H})}{C_\mathcal{I}(\mathcal{S}_\mathbf{X}, \mathcal{T}_\mathbf{X})(1-\varepsilon)\varepsilon}\right)\right],$$

and an i.i.d. sample T_u *from* $\mathcal{T}_\mathbf{X}$ *of size*

$$|T_u| \geq \frac{1}{\epsilon}\left(2\left[\frac{\sqrt{d}}{\gamma}\right]^d \ln\left(3\left[\frac{\sqrt{d}}{\gamma}\right]^d \delta\right)\right),$$

$\mathcal{A}$ *outputs a classifier* h *with* $\mathrm{R}_\mathcal{T}(h, f) \leq \epsilon$ *with probability at least* $1 - \delta$.

PROOF.– Let $\epsilon > 0$ and $\delta > 0$ be given and set $C = C_\mathcal{I}(\mathcal{S}_\mathbf{X}, \mathcal{T}_\mathbf{X})$. Set further $\epsilon' = \epsilon/2$ and $\delta' = \delta/3$ and divide the space $\mathbf{X}$ into a union of sets of heavy and light boxes from $\mathcal{B}$, where a box $b \in \mathcal{B}$ is called light if $\mathcal{T}_\mathbf{X}(b) \leq \epsilon'/|\mathcal{B}| = \epsilon'/(\sqrt{d}/\gamma)^d$ and heavy otherwise. Now let $\mathbf{X}^l$ denote the union of the light boxes and $\mathbf{X}^h$ the union of the heavy boxes. Furthermore, denote by $\mathcal{S}_{\mathbf{X}}^h$ and $\mathcal{T}_{\mathbf{X}}^h$ the restrictions of the source and target distributions to $\mathbf{X}^h$, such that $\mathcal{S}_{\mathbf{X}}^h(U) = \mathcal{S}_{\mathbf{X}}(U)/\mathcal{S}_{\mathbf{X}}(\mathbf{X}^h)$ and $\mathcal{T}_{\mathbf{X}}^h(U) = \mathcal{T}_{\mathbf{X}}(U)/\mathcal{T}_{\mathbf{X}}(\mathbf{X}^h)$ for all $U \subseteq \mathbf{X}^h$ and $\mathcal{S}_{\mathbf{X}}^h(U) = \mathcal{T}_{\mathbf{X}}^h(U) = 0$ for all $U \nsubseteq \mathbf{X}^h$. Since $|\mathcal{B}| = (\sqrt{d}/\gamma)^d$, we obtain $\mathcal{T}_{\mathbf{X}}(\mathbf{X}^h) \geq 1 - \epsilon'$ and thus, $\mathcal{S}_{\mathbf{X}}(\mathbf{X}^h) \geq C(1 - \epsilon')$.

The following parts of proof are built around the following two claims:

Claim 1. With probability at least $1 - \delta'$, an i.i.d. $\mathcal{T}_{\mathbf{X}}$ sample T with size as stated in the theorem hits every heavy box.

Claim 2. With probability at least $1 - 2\delta'$, the intersection of S and $\mathbf{X}^h$, where S is an i.i.d. $\mathcal{S}_{\mathbf{X}}$ sample which size as stated in the theorem, is an ϵ'-net for $\mathcal{H}\Delta\mathcal{H}$ with respect to $\mathcal{T}_{\mathbf{X}}^h$.

In order to understand why these two claims imply the claim of the theorem, denote by $S^h = S \cap \mathbf{X}^h$ the intersection of the source sample and the union of heavy boxes. By claim 1, T hits every heavy box with high probability, thus $S^h \subseteq S'$, where S' is the intersection of S with boxes that are hit by T according to the description of the algorithm $\mathcal{A}$. Consequently, one can apply claim 2 to ensure that if S^h is an ϵ'-net for $\mathcal{H}\Delta\mathcal{H}$ with respect to $\mathcal{T}_{\mathbf{X}}^h$ then so is S'. Hence, with probability at least $1 - 3\delta' = 1 - \delta$ the set S' is an ϵ'-net for $\mathcal{H}\Delta\mathcal{H}$ with respect to $\mathcal{T}_{\mathbf{X}}^h$. Now one can observe that an ϵ'-net for $\mathcal{H}\Delta\mathcal{H}$ with respect to $\mathcal{T}_{\mathbf{X}}^h$ is an ϵ-net with respect to $\mathcal{T}_{\mathbf{X}}$, as every set of $\mathcal{T}_{\mathbf{X}}$-weight at least ϵ has $\mathcal{T}_{\mathbf{X}}^h$ weight at least ϵ', following the definition of $\mathbf{X}^h$ and $\mathcal{T}_{\mathbf{X}}^h$.

Finally, one has to show that if S' is an ϵ-net for the set $\mathcal{H}\Delta\mathcal{H}$ of symmetric differences with respect to the target class, then the risk minimizing classifier from the target class has a target error at most ϵ. To proceed, let $h_{\mathcal{T}}^* \in \mathcal{H}$ denote the γ-margin classifier of zero target error. In this case, every box in $\mathcal{B}$ is labeled homogeneously with label 1 or label 0 by the labeling function f as f is a γ-margin classifier as well. Let $s \in S'$ be some picked point and $b_s \in \mathcal{B}$ be the box that contains s. Recall that $h_{\mathcal{T}}^*$ is a classifier with margin γ and b_s was hit by T by the definition of S'. This means that $\mathcal{T}_{\mathbf{X}}(b_s) > 0$ and b_s is labeled homogeneously by $h_{\mathcal{T}}^*$, which has zero target error. From this, one can deduce that this label coincides with the labeling provided by f, i.e. $h_{\mathcal{T}}^*(s) = f(s)$ for all $s \in S'$, which means that the empirical error of $h_{\mathcal{T}}^*$ with respect to S' is zero. Now consider a classifier h_ϵ with $\mathrm{R}_{\mathcal{T}}(h_\epsilon) \geq \epsilon$. Let $s \in S'$ be a point in $h_{\mathcal{T}}^* \Delta h_\epsilon$ whose existence is ensured by S' being an ϵ-net. As $s \in h_{\mathcal{T}}^T \Delta h_\epsilon$, we immediately obtain that $h_\epsilon(s) = h_{\mathcal{T}}^T(s) = f(s)$ and thus, h_ϵ has an empirical error larger than zero, which implies that no classifier of error larger than ϵ can be chosen by the risk minimizing algorithm on input S'.

Claim 1 can be now proved as follows. Let b be a heavy box implying $\mathcal{T}_{\mathbf{X}}(b) \geq \epsilon'/|\mathcal{B}|$. Then, for a sample T drawn i.i.d. from $\mathcal{T}_{\mathbf{X}}$, the probability of not hitting b is at most $(1-(\epsilon'/|\mathcal{B}|))^{|T|}$. Consequently, the union bound implies that the probability that there exists a box in $\mathcal{B}^h$ that does not get hit by the sample T is bounded by

$$\left|\mathcal{B}^h\right|(1-(\epsilon'/|\mathcal{B}|))^{|T|} \le |\mathcal{B}|(1-(\epsilon'/|\mathcal{B}|))^{|T|} \le |\mathcal{B}|\, e^{-\epsilon'|T|/|\mathcal{B}|)}.$$

Now if

$$|T| \ge \frac{|\mathcal{B}|\log(|\mathcal{B}|/\delta')}{\epsilon'} = \frac{2(\sqrt{d}/\gamma)^d \log(3(\sqrt{d}/\gamma)^d/\delta)}{\epsilon},$$

then the sample T hits every heavy box with probability at least $1-\delta'$ which proves claim 1.

We now proceed to the proof of claim 2. Let S^h be defined as $S \cap \mathbf{X}^h$. Note that, as S is an i.i.d. sample drawn from $\mathcal{S}_{\mathbf{X}}$, we can consider S^h to be an i.i.d. sample drawn from $\mathcal{S}^h_{\mathbf{X}}$. In this case, the following bound on the weight ratio between $\mathcal{S}^h_{\mathbf{X}}$ and $\mathcal{T}^h_{\mathbf{X}}$ holds:

$$\begin{aligned} C_{\mathcal{P}}(\mathcal{S}^h_{\mathbf{X}}, \mathcal{T}^h_{\mathbf{X}}) &= \inf_{p\in\mathcal{P}, \mathcal{T}^h_{\mathbf{X}}(p)>0} \frac{\mathcal{S}^h_{\mathbf{X}}(p)}{\mathcal{T}^h_{\mathbf{X}}(p)} = \inf_{p\in\mathcal{P}, \mathcal{T}^h_{\mathbf{X}}(p)>0} \frac{\mathcal{S}_{\mathbf{X}}(p)\mathcal{S}_{\mathbf{X}}(\mathbf{X}^h)}{\mathcal{T}_{\mathbf{X}}(p)\mathcal{T}_{\mathbf{X}}(\mathbf{X}^h)} \\ &\ge C\frac{\mathcal{S}_{\mathbf{X}}(\mathbf{X}^h)}{\mathcal{T}_{\mathbf{X}}(\mathbf{X}^h)} \ge C(1-\epsilon'), \end{aligned}$$

where the last inequality is obtained due to $\mathcal{T}_{\mathbf{X}}(\mathbf{X}^h) \ge C(1-\epsilon')$ and $\mathcal{S}_{\mathbf{X}}(\mathbf{X}^h) \le 1$. Note that every element in $\mathcal{H}\Delta\mathcal{H}$ can be partitioned into elements from $\mathcal{P}$, so that the same bound on the weight ratio can be obtained for the symmetric differences of $\mathcal{H}$ too:

$$C_{\mathcal{H}\Delta\mathcal{H}}(\mathcal{S}^h_{\mathbf{X}}, \mathcal{T}^h_{\mathbf{X}}) \ge C(1-\epsilon').$$

We can now state that there is a constant $c > 1$ such that if S^h has size at least

$$M; = c\left(\frac{VC(\mathcal{H}\Delta\mathcal{H}) + \log(1/\delta)}{C_{\mathcal{I}}(1-\epsilon')\epsilon'} \frac{VC(log(VC(\mathcal{H}\Delta\mathcal{H}))}{C_{\mathcal{I}}(1-\epsilon')\epsilon'}\right)$$

then, with probability at least $1-\delta'$, it is an $C_{\mathcal{I}}(1-\epsilon')\epsilon'$-net with respect to $\mathcal{S}^h_{\mathbf{X}}$ and thus an ϵ'-net with respect to $\mathcal{T}_{\mathbf{X}}(\mathbf{X}^h)$ (see, for example, corollary 3.8 in [BAT 13]). Thus, we only need to show that $\left|S^h\right| \ge M$ with probability at least $1-\delta'$. As $\mathcal{S}_{\mathbf{X}}(\mathbf{X}^h) \ge C(1-\epsilon')$, the sampling of the points of S and checking whether they hit

$\mathbf{X}^h$ can be seen as a Bernoulli random variable with mean $\mu = \mathcal{S}_{\mathbf{X}}(\mathbf{X}^h) \geq C(1-\epsilon')$. Using Hoeffding's inequality, we obtain that, $\forall t > 0$,

$$\mathbf{Pr}\left[\mu\,|S| - \left|S^h\right| \geq t\,|S|\right] \leq e^{-2t^2|S|}.$$

Setting $C' = C(1-\epsilon')$, $t = C'/2$ and assuming $|S| \geq 2M$ in the above inequality yields:

$$\mathbf{Pr}\left[\left|S^h\right| \leq M\right] \leq \mathbf{Pr}\left[\mu\,|S| - \left|S^h\right| \geq \frac{C'}{2}\,|S|\right] \leq e^{\frac{-C'^2|S|}{2}}.$$

From this,

$$|S| \geq \frac{2M}{C} > \frac{2(VC(\mathcal{H}\Delta\mathcal{H}) + \log(1/\delta'))}{C^2(1-\epsilon')^2\epsilon'}$$

implies that $e^{-\frac{C'^2|S|}{2}} \leq \delta'$, meaning in turn that S^h is an ϵ'-net of $\mathcal{H}\Delta\mathcal{H}$ with probability at least $1 - 2\delta'$. Bearing in mind that $VC(\mathcal{H}\Delta\mathcal{H}) \leq 2VC(\mathcal{H}) + 1$, it completes the proof. □

THEOREM 4.7 ([BEN 12b]).– *Let domain* $\mathbf{X} = [0,1]^d$ *and for some* $C > 0$, *let* $\mathcal{W}$ *be a class of pairs of source and target distributions* $\{(\mathcal{S},\mathcal{T})|C_{\mathcal{B}}(\mathcal{S}_{\mathbf{X}},\mathcal{T}_{\mathbf{X}}) \geq C\}$ *with bounded weight ratio and their common labeling function* $f : \mathbf{X} \to [0,1]$, *satisfying the* ϕ*-probabilistic-Lipschitz property with respect to the target distribution, for some function* ϕ. *Then, for all* λ,

$$\mathop{\mathbf{E}}_{S\sim\mathcal{S}^m}\left[\mathrm{R}_{\mathcal{T}}(h_{NN})\right] \leq 2\mathrm{R}^*_{\mathcal{T}}(\mathcal{H}) + \phi(\lambda) + 4\lambda\frac{\sqrt{d}}{C}m^{-\frac{1}{d-1}}.$$

PROOF.– The first intermediate result that we prove is the following:

$$\mathop{\mathbf{E}}_{S\sim\mathcal{S}^m}\left[\mathrm{R}_{\mathcal{T}}(h_{\mathrm{NN}})\right] \leq 2\mathrm{R}^*_{\mathcal{T}}(\mathcal{H}) + \phi(\lambda) + \lambda \mathop{\mathbf{E}}_{\substack{S\sim\mathcal{S}^m\\ \mathbf{x}\sim\mathcal{T}_{\mathbf{X}}}}\left[\|\mathbf{x} - N_S(\mathbf{x})\|_2\right]. \qquad \text{[A2.1]}$$

To proceed, we first note that given two instances $\mathbf{x}$, $\mathbf{x}'$ the following holds:

$$\begin{aligned}\mathop{\mathbf{Pr}}_{\substack{y\sim f(\mathbf{x})\\ y'\sim f(\mathbf{x}')}}[y \neq y'] &= f(\mathbf{x})(1-f(\mathbf{x}')) + f(\mathbf{x}')(1-f(\mathbf{x}))\\ &\leq 2f(\mathbf{x})(1-f(\mathbf{x})) + |f(\mathbf{x}') - f(\mathbf{x})|.\end{aligned}$$

Here, the last inequality is due to standard algebraic manipulations. From this, the left-hand side of the statement can be written as:

$$\begin{aligned}\mathop{\mathbf{E}}_{S\sim\mathcal{S}^m}[\mathrm{R}_{\mathcal{T}}(h_{\mathrm{NN}})] &= \mathop{\mathbf{E}}_{S\sim\mathcal{S}^m}\mathop{\mathbf{E}}_{\mathbf{x}\sim\mathcal{T}_{\mathbf{X}}}\mathop{\mathbf{Pr}}_{\substack{y\sim f(\mathbf{x})\\ y'\sim f(\mathbf{x}')}}[y\neq y'] \\ &\leq \mathop{\mathbf{E}}_{S\sim\mathcal{S}^m}\mathop{\mathbf{E}}_{\mathbf{x}\sim\mathcal{T}_{\mathbf{X}}}[2f(\mathbf{x})(1-f(\mathbf{x}))+|f(\mathbf{x}')-f(\mathbf{x})|] \\ &\leq 2\mathrm{R}^*_{\mathcal{T}}(\mathcal{H}) + \mathop{\mathbf{E}}_{S\sim\mathcal{S}^m}\mathop{\mathbf{E}}_{\mathbf{x}\sim\mathcal{T}_{\mathbf{X}}}[|f(N_S(\mathbf{x}'))-f(\mathbf{x})|].\end{aligned}$$

Furthermore, one can use the definition of probabilistic Lipschitzness and the fact that the range of f is $[0,1]$, regardless of the choice of S, to obtain:

$$\mathop{\mathbf{E}}_{\mathbf{x}\sim\mathcal{T}_{\mathbf{X}}}[|f(N_S(\mathbf{x}'))-f(\mathbf{x})|] \leq \phi(\lambda) + \lambda \mathop{\mathbf{E}}_{\mathbf{x}\sim\mathcal{T}_{\mathbf{X}}}[\|\mathbf{x}-N_S(\mathbf{x})\|_2],$$

which proves the inequality given in equation [A2.1]. To obtain the final result, one has to upper bound $\mathop{\mathbf{E}}_{S\sim\mathcal{S}^m}\mathop{\mathbf{E}}_{\mathbf{x}\sim\mathcal{T}_{\mathbf{X}}}[|f(N_S(\mathbf{x}'))-f(\mathbf{x})|]$.

To this end, we fix some $\gamma > 0$ and let $C_1, \ldots, C_r$ be the cover of the set $[0,1]^d$ using boxes of side-length γ. In this case, we obtain that for all boxes C_i, $\mathcal{T}_{\mathbf{X}}(C_i) \leq \frac{1}{C_{\mathcal{B}}(\mathcal{S}_{\mathbf{X}},\mathcal{T}_{\mathbf{X}})}\mathcal{S}_{\mathbf{X}}(C_i) \leq \frac{1}{C}\mathcal{S}_{\mathbf{X}}(C_i)$.

In order to proceed, we need the following lemma.

LEMMA A2.3.– *Let $C_1, C_2, \ldots, C_r$ be subsets of some domain set $\mathbf{X}$ and S be a set of points of size m sampled i.i.d. according to some distribution $\mathcal{D}_{\mathbf{X}}$ over that domain. Then,*

$$\mathop{\mathbf{E}}_{S\sim\mathcal{D}^m_{\mathbf{X}}}\Big[\sum_{i:C_i\cap S=\emptyset}\mathcal{D}_{\mathbf{X}}[C_i]\Big] \leq \frac{r}{me}.$$

We first use the linearity of expectation to obtain:

$$\mathop{\mathbf{E}}_{S\sim\mathcal{D}^m_{\mathbf{X}}}\Big[\sum_{i:C_i\cap S=\emptyset}\mathcal{D}_{\mathbf{X}}[C_i]\Big] = \sum_{i=1}^{r}\mathcal{D}_{\mathbf{X}}[C_i]\mathop{\mathbf{E}}_{S\sim\mathcal{D}^m_{\mathbf{X}}}[\mathbf{1}_{i:C_i\cap S=\emptyset}].$$

Then, for each i the following holds:

$$\begin{aligned}\mathop{\mathbf{E}}_{S\sim\mathcal{D}^m_{\mathbf{X}}}[\mathbf{1}_{i:C_i\cap S=\emptyset}] &= \mathop{\mathbf{Pr}}_{S\sim\mathcal{D}^m_{\mathbf{X}}}[C_i\cap S=\emptyset] \\ &= (1-\mathcal{D}_{\mathbf{X}}[C_i])^m \leq e^{-\mathcal{D}_{\mathbf{X}}[C_i])m}.\end{aligned}$$

The lemma is then proved by combining the above two equations with the fact that $\max_a ae^{-ma} \leq \frac{1}{me}$. We can now apply this lemma to get:

$$\underset{S\sim\mathcal{S}_{\mathbf{X}}^m}{\mathbf{E}}\Big[\sum_{i:C_i\cap S=\emptyset}\mathcal{T}_{\mathbf{X}}[C_i]\Big] = \underset{S\sim\mathcal{S}_{\mathbf{X}}^m}{\mathbf{E}}\Big[\sum_{i:C_i\cap S=\emptyset}\frac{1}{C}\mathcal{S}_{\mathbf{X}}[C_i]\Big]$$
$$\leq \frac{1}{C}\underset{S\sim\mathcal{S}_{\mathbf{X}}^m}{\mathbf{E}}\Big[\sum_{i:C_i\cap S=\emptyset}\mathcal{S}_{\mathbf{X}}[C_i]\Big] \leq \frac{r}{Cme}.$$

If $\mathbf{x},\mathbf{x}'$ are in the same box, we have that their distance is bounded as $\|\mathbf{x}-\mathbf{x}'\|^2 \leq \sqrt{d}\gamma$ while in the opposite case, $\|\mathbf{x}-\mathbf{x}'\|^2 \leq 2\sqrt{d}$. For $\mathbf{x}\in\mathbf{X}$, we let $C_x \in C_1,\ldots,C_r$ denote the box that contains the point $\mathbf{x}$. Then, we get:

$$\underset{\substack{S\sim\mathcal{S}^m\\ \mathbf{x}\sim\mathcal{T}_{\mathbf{X}}}}{\mathbf{E}}[\|\mathbf{x}-N_S(\mathbf{x})\|_2] \leq \underset{S\sim\mathcal{S}^m}{\mathbf{E}}\Big[\underset{\mathbf{x}\sim\mathcal{T}_{\mathbf{X}}}{\mathbf{Pr}}[C_x\cap S=\emptyset]2\sqrt{d}$$
$$+\underset{\mathbf{x}\sim\mathcal{T}_{\mathbf{X}}}{\mathbf{Pr}}[C_x\cap S=\emptyset]\sqrt{d}\gamma\Big] \leq \sqrt{d}\left(\frac{2r}{Cme}+\gamma\right).$$

Since the number of boxes is $(\frac{2}{\gamma})^d$, we immediately obtain

$$\underset{\substack{S\sim\mathcal{S}^m\\ \mathbf{x}\sim\mathcal{T}_{\mathbf{X}}}}{\mathbf{E}}[\|\mathbf{x}-N_S(\mathbf{x})\|_2] \leq \sqrt{d}\left(\frac{2^{d+1}\gamma^{-d}}{Cme}+\gamma\right).$$

This last result can be combined with equation [A2.1] to yield:

$$\underset{S\sim\mathcal{S}^m}{\mathbf{E}}[\mathrm{R}_{\mathcal{T}}(h_{\mathrm{NN}})] \leq 2\mathrm{R}^*_{\mathcal{T}}(\mathcal{H})+\phi(\lambda)+\lambda\sqrt{d}\left(\frac{2^{d+1}\gamma^{-d}}{Cme}+\gamma\right)$$
$$\leq 2\mathrm{R}^*_{\mathcal{T}}(\mathcal{H})+\phi(\lambda)+\lambda\sqrt{d}\frac{1}{C}\left(\frac{2^{d+1}\gamma^{-d}}{me}+\gamma\right)$$

where the last line is due to the fact that $0<C\leq 1$. Setting $\gamma = 2m^{-\frac{1}{d+1}}$ and noting that

$$\frac{2^{d+1}\gamma^{-d}}{Cme}+\gamma = \frac{2^{d+1}\gamma^{-d}m^{\frac{d}{d+1}}}{Cme}+2m^{-\frac{1}{d+1}}$$
$$=2m^{-\frac{1}{d+1}}\left(\frac{1}{e}+1\right) \leq 4m^{-\frac{1}{d+1}}$$

gives the final result. □

THEOREM 4.8 ([BEN 12b]).– *Let domain $\mathbf{X} = [0,1]^d$ for some d. Consider the class $\mathcal{H}$ of half-spaces as the target class. Let $\mathbf{x}$ and $\mathbf{z}$ be a pair of antipodal points on the unit sphere and $\mathcal{W}$ be a set that contains two pairs $(\mathcal{S},\mathcal{T})$ and $(\mathcal{S}',\mathcal{T}')$ of distributions where:*

1) both pairs satisfy the covariate shift assumption;

2) $f(\mathbf{x}) = f(\mathbf{z}) = 1$ *and* $f(\overline{0}) = 0$ *for their common labeling function f;*

3) $\mathcal{S}_{\mathbf{X}}(\mathbf{x}) = \mathcal{T}_{\mathbf{X}}(\mathbf{z}) = \mathcal{S}_{\mathbf{X}}(\overline{0}) = \frac{1}{3}$*;*

4) $\mathcal{T}_{\mathbf{X}}(\mathbf{x}) = \mathcal{T}_{\mathbf{X}}(\overline{0}) = \frac{1}{2}$ *or* $\mathcal{T}'_{\mathbf{X}}(\mathbf{z}) = \mathcal{T}'_{\mathbf{X}}(\overline{0}) = \frac{1}{2}$.

Then, for any number m*, any constant* c*, no proper DA learning algorithm can* $(c, \varepsilon, \delta, m, 0)$*-solve the domain adaptation learning task for* $\mathcal{W}$ *with respect to* $\mathcal{H}$*, if* $\varepsilon < \frac{1}{2}$ *and* $\delta < \frac{1}{2}$*. In other words, every learner that ignores unlabeled target data fails to produce a zero-risk hypothesis with respect to* $\mathcal{W}$*.*

PROOF.– Obviously, there is no halfspace that can allow us to correctly classify the three points $\mathbf{x}$, $\overline{0}$ and y. Furthermore, one may note that for any halfspace h, we have $\mathrm{R}_{\mathcal{T}_{\mathbf{X}}}(h) + \mathrm{R}_{\mathcal{T}'_{\mathbf{X}}}(h) \geq 1$, which implies $\mathrm{R}_{\mathcal{T}_{\mathbf{X}}}(h) \geq \frac{1}{2}$ or $\mathrm{R}_{\mathcal{T}'_{\mathbf{X}}}(h) \geq \frac{1}{2}$. Thus, for every learner, there exists a target distribution that can be either $\mathcal{T}_{\mathbf{X}}$ or $\mathcal{T}'_{\mathbf{X}}$, such that it returns a function that achieves an error of at least $\frac{1}{2}$ with probability at least $\frac{1}{2}$ over the choice of sample. The desired result is obtained due to the fact that the approximation error of the class of halfspaces for the target distributions is 0, and thus the statement of the theorem holds for any constant c. □

THEOREM 4.9 ([BEN 12b]).– *Let* $\mathbf{X}$ *be some domain and* $\mathcal{W}$ *be a class of pairs* $(\mathcal{S}, \mathcal{T})$ *of distributions over* $\mathbf{X} \times \{0,1\}$ *with* $\mathrm{R}_{\mathcal{T}}(\mathcal{H}) = 0$ *such that there is an algorithm* $\mathcal{A}$ *and functions* $m : (0,1)^2 \to \mathbb{N}$*,* $n : (0,1)^2 \to \mathbb{N}$ *such that* $\mathcal{A}$ $(0, \varepsilon, \delta, m(\varepsilon, \delta), n(\varepsilon, \delta))$*-solves the domain adaptation learning task for* $\mathcal{W}$ *for all* $\varepsilon, \delta > 0$*. Let* $\mathcal{H}$ *be some hypotheses class for which there exists an agnostic proper learner. Then, the* $\mathcal{H}$*-proper domain adaptation problem can be* $((0, \varepsilon, \delta, m(\varepsilon/3, \delta/2), n(\varepsilon/3, \delta/2)) + m'(\varepsilon/3, \delta/2))$*-solved with respect to the class* $\mathcal{W}$*, where* m' *is the sample complexity function for agnostically learning* $\mathcal{H}$*.*

PROOF.– Given the parameters ϵ and δ, let S be a sample of size at least $m(\epsilon/3, \delta/2)$ drawn from $\mathcal{S}$, and T be an unlabeled sample of size $n(\epsilon/3, \delta/2) + m'(\epsilon/3, \delta/2)$ drawn from $\mathcal{T}_{\mathbf{X}}$. Based on these, one can split the unlabeled sample into a sample T_1 of size $n(\epsilon/3, \delta/2)$ and T_2 of size $m'(\epsilon/3, \delta/2)$. After that, we can apply the hypotheses returned by the learner $\mathcal{A}$ when applied to S and T_1 in order to label all members of T_2, and then use the newly labeled T_2 as input to the agnostic proper learner for $\mathcal{H}$. The final result can be obtained due to the following lemma.

LEMMA A2.4.– *Let* $\mathcal{D}$ *be a distribution over* $\mathbf{X} \times \{0,1\}$*,* $f : \mathbf{X} \to \{0,1\}$ *be a function with* $\mathrm{R}_{\mathcal{D}}(f) \leq \epsilon_0$*,* $\mathcal{A}$ *be an agnostic learner for some hypothesis class* $\mathcal{H}$ *over* $\mathbf{X}$ *and* $m : (0,1)^2 \to \mathbb{N}$ *be a function such that* $\mathcal{A}$*, for all* $\epsilon, \delta > 0$*, is guaranteed to* $(\epsilon, \delta, m(\epsilon, \delta))$*-learn* $\mathcal{H}$*. Then, with probability at least* $(1 - \delta)$ *over an i.i.d. sample of size* $m(\epsilon, \delta)$ *from* $\mathcal{D}_{\mathbf{X}}$*'s marginal labeled by* f*,* $\mathcal{A}$ *outputs a hypothesis* h *with* $\mathrm{R}_{\mathcal{D}}(h) \leq \mathrm{R}^*_{\mathcal{D}_{\mathbf{X}}}(\mathcal{H}) + 2\epsilon_0 + \epsilon$.

Let $\mathcal{D}'$ be the distribution with the same marginal as $\mathcal{D}$ and let f be their common deterministic labeling function. Note that for the optimal hypothesis h^* in $\mathcal{H}$ with respect to $\mathcal{D}$, we have $\mathrm{R}_{\mathcal{D}'}(h^*) \leq \mathrm{R}^*_{\mathcal{D}_\mathbf{X}}(\mathcal{H}) + \epsilon_0$. This implies that when we use the sample S generated by $\mathcal{D}'$ to the agnostic learner, it returns an $h \in \mathcal{H}$ with $\mathrm{R}_{\mathcal{D}'}(h^*) \leq \mathrm{R}^*_{\mathcal{D}_\mathbf{X}}(\mathcal{H}) + \epsilon + \epsilon_0$ and this with probability at least $(1 - \delta)$. Consequently, it yields $\mathrm{R}_{\mathcal{D}}(h) \leq \mathrm{R}^*_{\mathcal{D}_\mathbf{X}}(\mathcal{H}) + 2\epsilon_0 + \epsilon$. The claimed performance of the output hypothesis thus follows from this lemma. □

Appendix 3

Proofs of the Main Results of Chapter 5

THEOREM (General form of theorem 5.1).– *For a labeling function* $f \in \mathcal{G}$*, let* $\mathcal{F} = \{(\mathbf{x}, y) \to \ell(f(\mathbf{x}), y)\}$ *be a class consisting of the bounded functions with the range* $[a, b]$*. Let* $S_k = \{(\mathbf{x}_1^{(k)}, y_1^{(k)}), \ldots, (\mathbf{x}_m^{(k)}, y_m^{(k)})\}$ *be a labeled sample drawn from* $\mathcal{S}_k$ *of size* N_k *for all* $k \in [1, \ldots, K]$*. Let* $w = (w_1, \ldots, w_K) \in [0,1]^K$ *with* $\sum_{k=1}^K w_k = 1$*. Then given any arbitrary* $\xi \geq D_{\mathcal{F}}^{\mathbf{w}}(\mathcal{S}, \mathcal{T})$*, for any* $\prod_{k=1}^K N_k \geq \frac{8(b-a)}{\xi'^2}$ *and any* $\epsilon > 0$*, with probability at least* $1 - \epsilon$*, the following holds:*

$$\sup_{f \in \mathcal{F}} \left| \sum_{i=1}^{K} w_k \mathrm{R}_{\hat{\mathcal{S}}_k}^{\ell} f - \mathrm{R}_{\mathcal{T}}^{\ell} f \right| \leq D_{\mathcal{F}}^{\mathbf{w}}(\mathcal{S}, \mathcal{T})$$

$$+ \left(\frac{\ln \mathcal{N}_1^{\mathbf{w}}(\xi'/8, \mathcal{F}, 2\sum_{k=1}^K N_k) - \ln(\epsilon/8)}{\frac{\prod_{k=1}^K N_K}{32(b-a)^2 \sum_{k=1}^K w_k^2 (\prod_{i \neq k} N_i)}} \right)^{\frac{1}{2}},$$

where $\xi' = \xi - D_{\mathcal{F}}^{\mathbf{w}}(\mathcal{S}, \mathcal{T})$ *and*

$$D_{\mathcal{F}}^{\mathbf{w}}(\mathcal{S}, \mathcal{T}) = \sum_{i=1}^{K} w_k D_{\mathcal{F}}(\mathcal{S}_i, \mathcal{T}).$$

Here, the quantity $\mathcal{N}_1^{\mathbf{w}}(\xi, \mathcal{F}, 2\sum_{k=1}^K N_k)$ stands for the uniform entropy number that is given by the following equation:

$$\mathcal{N}_1^{\mathbf{w}}(\xi, \mathcal{F}, 2\sum_{k=1}^{K} N_k) = \sup_{\{S^{2N_k}\}_{k=1}^K} \log \mathcal{N}(\xi, \mathcal{F}, \ell_1^{\mathbf{w}}(\{S^{2N_k}\}_{k=1}^K)),$$

where for all source samples S_k and their associated ghost samples $S_k' = \{(\mathbf{x}_1'^{(k)}, y_1'^{(k)})\}_{i=1}^{N_k}$ drawn from $\mathcal{S}_k$, $\forall k = 1, \ldots, K$ the quantity

$S^{2N_k} = \{S_k, S_k'\}$ and the metric $\ell_1^{\mathbf{w}}$ is a variation of the ℓ_1 metric defined for some $f \in \mathcal{F}$ based on the following norm:

$$\|f\|_{\ell_1^{\mathbf{w}}(\{S^{2N_k}\}_{k=1}^K)} = \sum_{k=1}^K \frac{w_k}{N_k} \sum_{i=1}^{N_k} \left(|f((\mathbf{x}_i^{(k)}, y_i^{(k)}))| + |f((\mathbf{x}_i'^{(k)}, y_i'^{(k)}))| \right).$$

Compared to the classical result under the assumption of the same distribution relating the true and the empirical risk defined with respect to the sample set, there is a discrepancy quantity $D_{\mathcal{F}}^{\mathbf{w}}(\mathcal{S}, \mathcal{T})$ that appears in the proposed bound. The proposed result coincides with the classical one if any source domain and the target domain match, i.e., $D_{\mathcal{F}}(\mathcal{S}_i, \mathcal{T}) = 0$ holds for any $1 \leq k \leq K$.

THEOREM (General form of theorem 5.2).– *For a labeling function $f \in \mathcal{G}$, let $\mathcal{F} = \{(\mathbf{x}, y) \rightarrow \ell(f(\mathbf{x}), y)\}$ be a loss function class consisting of the bounded functions with the range $[a, b]$ for a space of labeling functions $\mathcal{G}$. Let $w = (w_1, \ldots, w_K) \in [0,1]^K$ with $\sum_{k=1}^K w_k = 1$. If the following holds*

$$\lim_{N_1,\ldots,N_K \rightarrow \infty} \frac{\ln \mathcal{N}_1^{\mathbf{w}}(\xi'/8, \mathcal{F}, 2\sum_{k=1}^K N_k)}{\frac{\prod_{k=1}^K N_K}{32(b-a)^2 \sum_{k=1}^K w_k^2 (\prod_{i \neq k} N_i)}} < \infty,$$

with $\xi' = \xi - D_{\mathcal{F}}^{\mathbf{w}}(\mathcal{S}, \mathcal{T})$, then we have for any $\xi \geq D_{\mathcal{F}}^{\mathbf{w}}(\mathcal{S}, \mathcal{T})$,

$$\lim_{m \rightarrow \infty} \mathbf{Pr} \{ \sup_{f \in \mathcal{F}} \left| \sum_{i=1}^K w_k \mathrm{R}_{\hat{\mathcal{S}}_k}^{\ell} f - \mathrm{R}_{\mathcal{T}}^{\ell} f \right| > \xi \} = 0.$$

By the Cauchy–Schwarz inequality, setting $w_k = \frac{N_k}{\sum_{k=1}^K N_k}$ for all $1 \leq k \leq K$ minimizes the second term of the theorem presented above and leads to the following result:

$$\sup_{f \in \mathcal{F}} \left| \sum_{i=1}^K w_k \mathrm{R}_{\hat{\mathcal{S}}_k}^{\ell} f - \mathrm{R}_{\mathcal{T}}^{\ell} f \right| \leq \frac{\sum_{i=1}^K N_k D_{\mathcal{F}}(\mathcal{S}_i, \mathcal{T})}{\sum_{k=1}^K N_k} + \left(\frac{\ln \mathcal{N}_1^{\mathbf{w}}(\xi'/8, \mathcal{F}, 2\sum_{k=1}^K N_k) - \ln(\epsilon/8)}{\frac{\sum_{k=1}^K N_K}{32(b-a)^2}} \right)^{\frac{1}{2}}.$$

This particular value for the weight vector w leads to the fastest rate of convergence that is confirmed by the numerical results provided by the authors in the original paper.

LEMMA 5.2 ([RED 17]).– *Let $\mathcal{S}_{\mathbf{X}}, \mathcal{T}_{\mathbf{X}} \in \mathcal{P}(\mathbf{X})$ be two probability measures on $\mathbb{R}^d$. Assume that the cost function $c(\mathbf{x}, \mathbf{x}') = \|\phi(\mathbf{x}) - \phi(\mathbf{x}')\|_{\mathcal{H}_{k_\ell}}$, where $\mathcal{H}$ is an RKHS equipped with kernel $k_\ell : \mathbf{X} \times \mathbf{X} \rightarrow \mathbb{R}$ induced by $\phi : \mathbf{X} \rightarrow \mathcal{H}_{k_\ell}$ and*

$k_\ell(\mathbf{x},\mathbf{x}') = \langle\phi(\mathbf{x}),\phi(\mathbf{x}')\rangle_{\mathcal{H}_{k_\ell}}$. *Assume further that the loss function* $\ell_{h,f} : \mathbf{x} \longrightarrow \ell(h(\mathbf{x}), f(\mathbf{x}))$ *is convex, symmetric and bounded and obeys the triangular equality and has the parametric form* $|h(\mathbf{x}) - f(\mathbf{x})|^q$ *for some* $q > 0$. *Assume also that kernel* k_ℓ *in the RKHS* $\mathcal{H}_{k_\ell}$ *is square-root integrable w.r.t. both* $\mathcal{S}_\mathbf{X}, \mathcal{T}_\mathbf{X}$ *for all* $\mathcal{S}_\mathbf{X}, \mathcal{T}_\mathbf{X} \in \mathcal{P}(\mathbf{X})$ *where* $\mathbf{X}$ *is separable and* $0 \leq k_\ell(\mathbf{x},\mathbf{x}') \leq K, \forall\ \mathbf{x},\mathbf{x}' \in \mathbf{X}$. *If* $\|\ell\|_{\mathcal{H}_{k_\ell}} \leq 1$, *then the following holds:*

$$\forall (h,h') \in \mathcal{H}_{k_\ell}^2, \quad \mathrm{R}_\mathcal{T}^{\ell_q}(h,h') \leq \mathrm{R}_\mathcal{S}^{\ell_q}(h,h') + W_1(\mathcal{S}_\mathbf{X},\mathcal{T}_\mathbf{X}).$$

PROOF.– We have

$$\begin{aligned}
\mathrm{R}_\mathcal{T}^{\ell_q}(h,h') &= \mathrm{R}_\mathcal{T}^{\ell_q}(h,h') + \mathrm{R}_\mathcal{S}^{\ell_q}(h,h') - \mathrm{R}_\mathcal{S}^{\ell_q}(h,h') \\
&= \mathrm{R}_\mathcal{S}^{\ell_q}(h,h') + \mathop{\mathbf{E}}_{\mathbf{x}'\sim\mathcal{T}_\mathbf{X}}[\langle\phi(\mathbf{x}'),\ell\rangle_{\mathcal{H}_{k_\ell}}] - \mathop{\mathbf{E}}_{\mathbf{x}\sim\mathcal{S}_\mathbf{X}}[\langle\phi(\mathbf{x}),\ell\rangle_{\mathcal{H}_{k_\ell}}] \\
&= \mathrm{R}_\mathcal{S}^{\ell_q}(h,h') + \langle \mathop{\mathbf{E}}_{\mathbf{x}'\sim\mathcal{T}_\mathbf{X}}[\phi(\mathbf{x}')] - \mathop{\mathbf{E}}_{\mathbf{x}\sim\mathcal{S}_\mathbf{X}}[\phi(\mathbf{x})],\ell\rangle_{\mathcal{H}_{k_\ell}} \\
&\leq \mathrm{R}_\mathcal{S}^{\ell_q}(h,h') + \|\ell\|_{\mathcal{H}_{k_\ell}} \| \mathop{\mathbf{E}}_{\mathbf{x}'\sim\mathcal{T}_\mathbf{X}}[\phi(\mathbf{x}')] - \mathop{\mathbf{E}}_{\mathbf{x}\sim\mathcal{S}_\mathbf{X}}[\phi(\mathbf{x})]\|_{\mathcal{H}_{k_\ell}} \\
&\leq \mathrm{R}_\mathcal{S}^{\ell_q}(h,h') + \|\int_\mathbf{X} \phi d(\mathcal{S}_\mathbf{X} - \mathcal{T}_\mathbf{X})\|_{\mathcal{H}_{k_\ell}}.
\end{aligned}$$

The second line is obtained by using the reproducing property applied to ℓ, and the third line follows from the properties of the expected value. The fourth line here is due to the properties of the inner product, whereas the fifth line is due to $\|\ell_{h,f}\|_{\mathcal{H}_{k_\ell}} \leq 1$. Now, using the dual form of the integral given by the Kantorovich–Rubinstein theorem, we have the following:

$$\begin{aligned}
\|\int_\mathbf{X} \phi d(\mathcal{S}_\mathbf{X} - \mathcal{T}_\mathbf{X})\|_{\mathcal{H}_{k_\ell}} &= \|\int_{\mathbf{X}\times\mathbf{X}} (\phi(\mathbf{x}) - \phi(\mathbf{x}'))d\gamma(\mathbf{x},\mathbf{x}')\|_{\mathcal{H}_{k_\ell}} \\
&\leq \int_{\mathbf{X}\times\mathbf{X}} \|\phi(\mathbf{x}) - \phi(\mathbf{x}')\|_{\mathcal{H}_{k_\ell}} d\gamma(\mathbf{x},\mathbf{x}').
\end{aligned}$$

Now we can obtain the final result by taking the infimum over γ from the left- and right-hand sides of the inequality, i.e.

$$\begin{aligned}
&\inf_{\gamma\in\Pi(\mathcal{S}_\mathbf{X},\mathcal{T}_\mathbf{X})} \|\int_\mathbf{X} \phi d(\mathcal{S}_\mathbf{X} - \mathcal{T}_\mathbf{X})\|_{\mathcal{H}_{k_\ell}} \\
&\leq \inf_{\gamma\in\Pi(\mathcal{S}_\mathbf{X},\mathcal{T}_\mathbf{X})} \int_{\mathbf{X}\times\mathbf{X}} \|\phi(\mathbf{x}) - \phi(\mathbf{x}')\|_{\mathcal{H}_{k_\ell}} d\gamma(\mathbf{x},\mathbf{x}').
\end{aligned}$$

which gives

$$\mathrm{R}_\mathcal{T}^{\ell_q}(h,h') \leq \mathrm{R}_\mathcal{S}^{\ell_q}(h,h') + W_1(\mathcal{S}_\mathbf{X},\mathcal{T}_\mathbf{X}).$$

□

THEOREM 5.5 ([RED 17]).– *Let S_u, T_u be unlabeled samples of size N_S and N_T each, drawn independently from $\mathcal{S}_{\mathbf{X}}$ and $\mathcal{T}_{\mathbf{X}}$, respectively. Let S be a labeled sample of size m generated by drawing $\beta\, m$ points from $\mathcal{T}_{\mathbf{X}}$ ($\beta \in [0,1]$) and $(1-\beta)\,m$ points from $\mathcal{S}_{\mathbf{X}}$ and labeling them according to f_S and f_T, respectively. If $\hat{h} \in \mathcal{H}$ is the empirical minimizer of $\mathrm{R}^{\alpha}_{\hat{S}}(h)$ on S and $h^*_T = \underset{h\in\mathcal{H}}{\operatorname{argmin}}\ \mathrm{R}^{\ell_q}_{\mathcal{T}}(h)$, then for any $\delta \in (0,1)$, with probability at least $1-\delta$ (over the choice of samples), we have*

$$\mathrm{R}^{\ell_q}_{\mathcal{T}}(\hat{h}) \ \leq\ \mathrm{R}^{\ell_q}_{\mathcal{T}}(h^*_T) + c_1 + 2(1-\alpha)(W_1(\hat{\mathcal{S}}_{\mathbf{X}}, \hat{\mathcal{T}}_{\mathbf{X}}) + \lambda + c_2),$$

where

$$c_1 = 2\sqrt{\frac{2K\left(\frac{(1-\alpha)^2}{1-\beta} + \frac{\alpha^2}{\beta}\right)\log(2/\delta)}{m}} + 4\sqrt{K/m}\left(\frac{\alpha}{m\beta\sqrt{\beta}} + \frac{(1-\alpha)}{m(1-\beta)\sqrt{1-\beta}}\right),$$

$$c_2 = \sqrt{2\log\left(\frac{1}{\delta}\right)/\varsigma'}\left(\sqrt{\frac{1}{N_S}} + \sqrt{\frac{1}{N_T}}\right).$$

PROOF.– The proof of this theorem follows the proof of theorem 3.3 and differs only in the concentration inequality obtained for the empirical combined error. This inequality is established by the following lemma.

LEMMA A3.1.– *Under the assumptions of lemma 5.2, let S be a labeled sample of size m generated by drawing $\beta\, m$ points from $\mathcal{T}_{\mathbf{X}}$ ($\beta \in [0,1]$) and $(1-\beta)\,m$ points from $\mathcal{S}_{\mathbf{X}}$ and labeling them according to f_S and f_T, respectively. Then with probability at least $1-\delta$ for all $h \in \mathcal{H}$ with $0 \leq k(x_i, x_j) \leq K$, the following holds:*

$$\mathbf{Pr}\left\{\left|\hat{\mathrm{R}}^{\alpha}(h) - \mathrm{R}^{\alpha}(h)\right| > 2\sqrt{K/m}\left(\frac{\alpha}{m\beta\sqrt{\beta}} + \frac{(1-\alpha)}{m(1-\beta)\sqrt{1-\beta}}\right) + \epsilon\right\}$$

$$\leq \exp\left\{\frac{-\epsilon^2 m}{2K\left(\frac{(1-\alpha)^2}{1-\beta} + \frac{\alpha^2}{\beta}\right)}\right\}.$$

PROOF.– First, we use McDiarmid's inequality in order to obtain the right side of the inequality by defining the maximum changes of magnitude when one of the sample vectors has been changed. □

For the sake of completeness, we give its definition here.

DEFINITION A3.1.– *Suppose $X_1, X_2, \ldots, X_n$ are independent random variables taking values in a set $\mathcal{X}$ and assume that $f : \mathcal{X}^n \to \mathbb{R}$ is a function of $X_1, X_2, \ldots, X_n$ that satisfies for $\forall i, \mathbf{x}_1, \ldots, \mathbf{x}_n, \mathbf{x}'_i \in \mathcal{X}$:*

$$|f(\mathbf{x}_1, \ldots, \mathbf{x}_i, \ldots, \mathbf{x}_n) - f(\mathbf{x}_1, \ldots, \mathbf{x}'_i, \ldots, \mathbf{x}_n)| \leq c_i.$$

Then the following inequality holds for any $\varepsilon > 0$:

$$P\left\{|f(x_1, x_2, \ldots, x_n) - \mathbf{E}\ [f(x_1, x_2, \ldots, x_n)]\,| > \varepsilon\right\} \leq \exp\left\{\frac{-2\epsilon^2}{\sum_{i=1}^{n} c_i^2}\right\}.$$

We now rewrite the difference between the empirical and true combined error in the following way:

$$|\hat{\mathrm{R}}^\alpha(h) - \mathrm{R}^\alpha(h)| = |\alpha(\mathrm{R}_{\mathcal{T}}^{\ell_q}(h) - \mathrm{R}_{\hat{\mathcal{T}}}^{\ell_q}(h)) - (\alpha - 1)(\mathrm{R}_{\mathcal{S}}^{\ell_q}(h) - \mathrm{R}_{\hat{\mathcal{S}}}^{\ell_q}(h))|$$

$$= |\alpha \mathop{\mathbf{E}}_{\mathbf{x}\sim\mathcal{T}_\mathbf{X}} [\ell_q(h(\mathbf{x}), f_{\mathcal{T}}(\mathbf{x}))] - (\alpha - 1) \mathop{\mathbf{E}}_{\mathbf{y}\sim\mathcal{S}_\mathbf{X}} [\ell_q(h(\mathbf{y}), f_{\mathcal{S}}(\mathbf{y}))]$$

$$- \frac{\alpha}{m\beta}\sum_{i=1}^{\beta m} \ell_q(h(\mathbf{x}_i), f_{\mathcal{T}}(\mathbf{x}_i)) + \frac{(\alpha - 1)}{m(1-\beta)}\sum_{i=1}^{m(1-\beta)} \ell_q(h(\mathbf{y}_i), f_{\mathcal{S}}(\mathbf{y}_i))|$$

$$\leq \sup_{\ell\in\mathcal{H}} |\alpha \mathop{\mathbf{E}}_{\mathbf{x}\sim\mathcal{T}_\mathbf{X}} [\ell_q(h(\mathbf{x}), f(\mathbf{x}))] - (\alpha - 1) \mathop{\mathbf{E}}_{\mathbf{y}\sim\mathcal{S}_\mathbf{X}} [\ell_q(h(\mathbf{y}), f(\mathbf{y}))]$$

$$- \frac{\alpha}{m\beta}\sum_{i=1}^{m\beta} \ell_q(h(\mathbf{x}_i), f_{\mathcal{T}}(\mathbf{x}_i)) + \frac{(\alpha - 1)}{m(1-\beta)}\sum_{i=1}^{m(1-\beta)} \ell_q(h(\mathbf{y}_i), f_{\mathcal{S}}(\mathbf{y}_i))|.$$

Changing either $\mathbf{x}_i$ or $\mathbf{y}_i$ in this expression changes its value by at most $\frac{2\alpha\sqrt{K}}{\beta m}$ and $\frac{2(1-\alpha)\sqrt{K}}{(1-\beta)m}$, respectively. This gives us the denominator of the exponential in definition A3.1 (see previous page)

$$\beta m\left(\frac{2\alpha\sqrt{K}}{\beta m}\right)^2 + (1-\beta)m\left(\frac{2(1-\alpha)\sqrt{K}}{(1-\beta)m}\right)^2 = \frac{4K}{m}\left(\frac{\alpha^2}{\beta} + \frac{(1-\alpha)^2}{(1-\beta)}\right).$$

Then, we bound the expectation of the difference between the true and empirical combined errors by the sum of Rademacher averages over the samples. Denoting by X' an i.i.d sample of size βm drawn independently of X (and likewise for Y'), and using the symmetrization technique we have

$$\mathop{\mathbf{E}}_{X,Y} \sup_{h\in\mathcal{H}} |\alpha \mathop{\mathbf{E}}_{\mathbf{x}\sim\mathcal{T}_\mathbf{X}} [\ell(h(\mathbf{x}), f(\mathbf{x}))] - (\alpha - 1) \mathop{\mathbf{E}}_{\mathbf{y}\sim\mathcal{S}_\mathbf{X}} [\ell(h(\mathbf{y}), f(\mathbf{y}))]$$

$$- \frac{\alpha}{m\beta}\sum_{i=1}^{m\beta} \ell(h(\mathbf{x}_i), f_{\mathcal{S}}(\mathbf{x}_i)) + \frac{(\alpha - 1)}{m(1-\beta)}\sum_{i=1}^{m(1-\beta)} \ell(h(\mathbf{y}_i), f_{\mathcal{T}}(\mathbf{y}_i))|$$

$$\leq \mathop{\mathbf{E}}_{X,Y} \sup_{h\in\mathcal{H}} |\mathop{\mathbf{E}}_{X'}\left(\frac{\alpha}{m\beta}\sum_{i=1}^{m\beta} \ell(h(\mathbf{x}'_i), f_{\mathcal{S}}(\mathbf{x}'_i))\right)$$

$$- (\alpha - 1)\mathop{\mathbf{E}}_{Y'}\left(\frac{(\alpha - 1)}{m(1-\beta)}\sum_{i=1}^{m\beta} \ell(h(\mathbf{y}'_i), f_{\mathcal{T}}(\mathbf{y}'_i))\right)$$

$$
\begin{aligned}
&- \frac{\alpha}{m\beta}\sum_{i=1}^{\beta m} \ell(h(\mathbf{x}_i), f_S(\mathbf{x}_i)) + \frac{(\alpha - 1)}{m(1-\beta)} \sum_{i=1}^{(1-\beta)m} \ell(h(\mathbf{y}_i), f_T(\mathbf{y}_i))| \\
&\leq \mathop{\mathbf{E}}_{X,X',Y,Y'} \sup_{h\in\mathcal{H}} |\frac{\alpha}{m\beta}\sum_{i=1}^{\beta m} \sigma_i(\ell(h(\mathbf{x}'_i), f_S(\mathbf{x}'_i)) - \ell(h(\mathbf{x}_i), f_S(\mathbf{x}_i))) \\
&+ \frac{1-\alpha}{m(1-\beta)}\sum_{i=1}^{\beta m} \sigma_i(\ell(h(\mathbf{y}'_i), f_T(\mathbf{y}'_i)) - \ell(h(\mathbf{y}_i), f_T(\mathbf{y}_i)))| \\
&\leq 2\sqrt{K/m}\left(\frac{\alpha}{m\beta\sqrt{\beta}} + \frac{(1-\alpha)}{m(1-\beta)\sqrt{1-\beta}}\right).
\end{aligned}
$$

Finally, the Rademacher averages, in their turn, are bounded using a theorem from [BAR 02]. Using this inequality in Definition A3.1 gives us the desired result:

$$
\begin{aligned}
&\mathbf{Pr}\left\{\left|\hat{\mathrm{R}}^{\alpha}(h) - \mathrm{R}^{\alpha}(h)\right| > 2\sqrt{K/m}\left(\frac{\alpha}{m\beta\sqrt{\beta}} + \frac{(1-\alpha)}{m(1-\beta)\sqrt{1-\beta}}\right) + \epsilon\right\} \\
&\leq \exp\left\{\frac{-\epsilon^2 m}{2K\left(\frac{(1-\alpha)^2}{1-\beta} + \frac{\alpha^2}{\beta}\right)}\right\}.
\end{aligned}
$$

The final result can now be obtained by following the steps in the proof of theorem 3.3 and applying the corresponding results with the Wasserstein distance and the lemma proved above. □

Appendix 4

Proofs of the Main Results of Chapter 6

THEOREM 6.1 ([GER 13]).– *Let $\mathcal{H}$ be a hypothesis class. We have*

$$\forall \rho \text{ on } \mathcal{H},\ \mathop{\mathbf{E}}_{h\sim\rho} \mathrm{R}_{\mathcal{T}}^{\ell_{01}}(h) \ \le\ \mathop{\mathbf{E}}_{h\sim\rho} \mathrm{R}_{\mathcal{S}}^{\ell_{01}}(h) + \frac{1}{2}\,\mathrm{dis}_\rho(\mathcal{S}_{\mathbf{X}}, \mathcal{T}_{\mathbf{X}}) + \lambda_\rho,$$

where λ_ρ is the deviation between the expected joint errors between pairs for voters on the target and source domains, which is defined as

$$\lambda_\rho \ =\ \Big| e_{\mathcal{T}}(\rho) - e_{\mathcal{S}}(\rho) \Big|. \qquad [6.3]$$

PROOF.– First, from equation [6.2], we recall that, given a domain $\mathcal{D}$ on $\mathbf{X} \times Y$ and a distribution ρ over $\mathcal{H}$, we have

$$\mathop{\mathbf{E}}_{h\sim\rho} \mathrm{R}_{\mathcal{D}}^{\ell_{01}}(h) \ =\ \frac{1}{2}\, d_{\mathcal{D}_{\mathbf{X}}}(\rho) + e_{\mathcal{D}}(\rho).$$

Therefore, we have

$$\begin{aligned}
\mathop{\mathbf{E}}_{h\sim\rho} \mathrm{R}_{\mathcal{T}}^{\ell_{01}}(h) - \mathop{\mathbf{E}}_{h\sim\rho} \mathrm{R}_{\mathcal{S}}^{\ell_{01}}(h) \ &=\ \frac{1}{2}\Big(d_{\mathcal{T}_{\mathbf{X}}}(\rho) - d_{\mathcal{S}_{\mathbf{X}}}(\rho)\Big) + \Big(e_{\mathcal{T}}(\rho) - e_{\mathcal{S}}(\rho)\Big) \\
&\le\ \frac{1}{2}\Big|d_{\mathcal{T}_{\mathbf{X}}}(\rho) - d_{\mathcal{S}_{\mathbf{X}}}(\rho)\Big| + \Big|e_{\mathcal{T}}(\rho) - e_{\mathcal{S}}(\rho)\Big| \\
&=\ \frac{1}{2}\,\mathrm{dis}_\rho(\mathcal{S}_{\mathbf{X}}, \mathcal{T}_{\mathbf{X}}) + \lambda_\rho.
\end{aligned}$$

□

THEOREM 6.3 ([GER 16]).– *Let $\mathcal{H}$ be a hypothesis space, let $\mathcal{S}$ and $\mathcal{T}$, respectively, be the source and the target domains on $\mathbf{X} \times Y$ and let $q > 0$ be a constant. We have for all posterior distributions ρ on $\mathcal{H}$,*

$$\mathop{\mathbf{E}}_{h\sim\rho} \mathrm{R}_{\mathcal{T}}^{\ell_{01}}(h) \leq \frac{1}{2} d_{\mathcal{T}_\mathbf{X}}(\rho) + \beta_q \times \Big[e_{\mathcal{S}}(\rho)\Big]^{1-\frac{1}{q}} + \eta_{\mathcal{T}\setminus\mathcal{S}},$$

where

$$\eta_{\mathcal{T}\setminus\mathcal{S}} = \mathop{\mathbf{Pr}}_{(\mathbf{x},y)\sim\mathcal{T}} \Big((\mathbf{x},y) \notin \mathrm{SUPP}(\mathcal{S})\Big) \sup_{h\in\mathcal{H}} \mathrm{R}_{\mathcal{T}\setminus\mathcal{S}}(h)$$

with $\mathcal{T}\setminus\mathcal{S}$ the distribution of $(\mathbf{x},y)\sim\mathcal{T}$ conditional an $(\mathbf{x},y) \in \mathrm{SUPP}(\mathcal{T})\setminus\mathrm{SUPP}(\mathcal{S})$.

PROOF.– From equation [6.2], we know that

$$\mathop{\mathbf{E}}_{h\sim\rho} \mathrm{R}_{\mathcal{T}}^{\ell_{01}}(h) = \tfrac{1}{2} d_{\mathcal{T}_\mathbf{X}}(\rho) + e_{\mathcal{T}}(\rho).$$

Let us split $e_{\mathcal{T}}(\rho)$ into two parts:

$$\begin{aligned} e_{\mathcal{T}}(\rho) &= \mathop{\mathbf{E}}_{(\mathbf{x},y)\sim\mathcal{T}} \mathop{\mathbf{E}}_{(h,h')\sim\rho^2} \ell_{01}(h(\mathbf{x}),y)\, \ell_{01}(h'(\mathbf{x}),y) \\ &= \mathop{\mathbf{E}}_{(\mathbf{x},y)\sim\mathcal{S}} \frac{\mathcal{T}(\mathbf{x},y)}{\mathcal{S}(\mathbf{x},y)} \mathop{\mathbf{E}}_{(h,h')\sim\rho^2} \ell_{01}(h(\mathbf{x}),y)\, \ell_{01}(h'(\mathbf{x}),y) \qquad \text{[A4.1]} \\ &\quad + \mathop{\mathbf{E}}_{(\mathbf{x},y)\sim\mathcal{T}} \mathbf{I}\left[(\mathbf{x},y) \notin \mathrm{SUPP}(\mathcal{S})\right] \mathop{\mathbf{E}}_{(h,h')\sim\rho^2} \ell_{01}(h(\mathbf{x}),y)\, \ell_{01}(h'(\mathbf{x}),y). \qquad \text{[A4.2]} \end{aligned}$$

(i) On the one hand, we upper bound the first part (line A4.1) using Hölder's inequality, with p such that $\frac{1}{p} = 1-\frac{1}{q}$:

$$\begin{aligned} &\mathop{\mathbf{E}}_{(\mathbf{x},y)\sim\mathcal{S}} \frac{\mathcal{T}(\mathbf{x},y)}{\mathcal{S}(\mathbf{x},y)} \mathop{\mathbf{E}}_{(h,h')\sim\rho^2} \ell_{01}(h(\mathbf{x}),y)\, \ell_{01}(h'(\mathbf{x}),y) \\ &\leq \left[\mathop{\mathbf{E}}_{(\mathbf{x},y)\sim\mathcal{S}} \left(\frac{\mathcal{T}(\mathbf{x},y)}{\mathcal{S}(\mathbf{x},y)}\right)^q\right]^{\frac{1}{q}} \left[\mathop{\mathbf{E}}_{(h,h')\sim\rho^2} \mathop{\mathbf{E}}_{(\mathbf{x},y)\sim\mathcal{S}} \left[\ell_{01}(h(\mathbf{x}),y)\, \ell_{01}(h'(\mathbf{x}),y)\right]^p\right]^{\frac{1}{p}} \\ &= \beta_q \times \Big[e_{\mathcal{S}}(\rho)\Big]^{\frac{1}{p}}, \end{aligned}$$

where we have removed the exponent from expression $[\ell_{01}(h(\mathbf{x}),y)\ell_{01}(h'(\mathbf{x}),y)]^p$ without affecting its value, which is either 1 or 0.

(ii) On the other hand, we upper bound the second part (line A4.2) by the term $\eta_{\mathcal{T}\setminus\mathcal{S}}$;

$$
\begin{aligned}
&\mathop{\mathbf{E}}_{(\mathbf{x},y)\sim\mathcal{T}} \left(\mathbf{I}\left[(\mathbf{x},y) \notin \text{SUPP}(\mathcal{S})\right] \mathop{\mathbf{E}}_{(h,h')\sim\rho^2} \ell_{01}\left(h(\mathbf{x}),y\right) \ell_{01}\left(h'(\mathbf{x}),y\right) \right) \\
&= \left(\mathop{\mathbf{E}}_{(\mathbf{x},y)\sim\mathcal{T}} \mathbf{I}\left[(\mathbf{x},y) \notin \text{SUPP}(\mathcal{S})\right] \right) \mathop{\mathbf{E}}_{(\mathbf{x},y)\sim\mathcal{T}\setminus\mathcal{S}} \mathop{\mathbf{E}}_{(h,h')\sim\rho^2} \ell_{01}\left(h(\mathbf{x}),y\right) \ell_{01}\left(h'(\mathbf{x}),y\right) \\
&= \left(\mathop{\mathbf{E}}_{(\mathbf{x},y)\sim\mathcal{T}} \mathbf{I}\left[(\mathbf{x},y) \notin \text{SUPP}(\mathcal{S})\right] \right) e_{\mathcal{T}\setminus\mathcal{S}}(\rho) \\
&= \left(\mathop{\mathbf{E}}_{(\mathbf{x},y)\sim\mathcal{T}} \mathbf{I}\left[(\mathbf{x},y) \notin \text{SUPP}(\mathcal{S})\right] \right) \left(\mathop{\mathbf{E}}_{h\sim\rho} \mathrm{R}^{\ell_{01}}_{\mathcal{T}\setminus\mathcal{S}}(h) - \tfrac{1}{2} d_{\mathcal{T}\setminus\mathcal{S}}(\rho) \right) \\
&\leq \left(\mathop{\mathbf{E}}_{(\mathbf{x},y)\sim\mathcal{T}} \mathbf{I}\left[(\mathbf{x},y) \notin \text{SUPP}(\mathcal{S})\right] \right) \sup_{h\in\mathcal{H}} \mathrm{R}^{\ell_{01}}_{\mathcal{T}\setminus\mathcal{S}}(h) \; = \; \eta_{\mathcal{T}\setminus\mathcal{S}}.
\end{aligned}
$$

□

Appendix 5

Proofs of the Main Results of Chapter 8

THEOREM 8.1 ([HAB 13b]).– *Let $h^{(i)}$ be a weak hypothesis learned at step i from $S^{(i)}$, and $\tilde{\mathrm{R}}_S^{(i)}(h^{(i)}) = \frac{1}{2} - \gamma_S^{(i)}$ its corresponding empirical error*[1]. *Let $\mathrm{R}_{\hat{T}}(h^{(i)}) = \frac{1}{2} - \gamma_T^{(i)}$ be the (unknown) empirical error of $h^{(i)}$ over the target sample T_u. $h^{(i)}$ is a weak domain adaptation learner w.r.t. a set $SL_j = \{\mathbf{x}_1^T, \ldots, \mathbf{x}_{2k}^T\}$ of $2k$ pseudo-labeled target data inserted at step j ($j \leq i$) if $\gamma_T^{(j-1)} > 0$.*

PROOF.– A pseudo-labeled point inserted at time j is mislabeled by $h^{(i)}$ either if it has a wrong pseudo-label (given by $h^{(j-1)}$ with probability $(\frac{1}{2} - \gamma_T^{(j-1)})$) and has been correctly classified by $h^{(i)}$ (with probability $(\frac{1}{2} + \gamma_S^{(i)})$) or if it has a correct pseudo-label (given by $h^{(j-1)}$ with probability $(\frac{1}{2} + \gamma_T^{(j-1)})$) and has been misclassified by $h^{(i)}$ (with probability $(\frac{1}{2} - \gamma_S^{(i)})$). Following definition 8.3, $h^{(i)}$ is a weak domain adaptation learner w.r.t. a set $SL_j = \{\mathbf{x}_1^T, \ldots, \mathbf{x}_{2k}^T\}$ of $2k$ pseudo-labeled target data inserted at step j if

$$\begin{aligned}
&\hat{\mathrm{R}}^{(j)}_{\mathcal{S}_{2k}}(h^{(i)}) = \mathbb{E}_{\mathbf{x}_l^T \in SL_j}[h^{(i)}(\mathbf{x}_l^T) \neq y_l^T] < \frac{1}{2} \\
\Leftrightarrow\quad &\left(\frac{1}{2} + \gamma_S^{(i)}\right)\left(\frac{1}{2} - \gamma_T^{(j-1)}\right) + \left(\frac{1}{2} - \gamma_S^{(i)}\right)\left(\frac{1}{2} + \gamma_T^{(j-1)}\right) < \frac{1}{2} \\
\Leftrightarrow\quad &\gamma_S^{(i)}\gamma_T^{(j-1)} > 0
\end{aligned}$$

1 In this self-labeling domain adaptation context, the tilde symbol ($\tilde{}$) is used instead of the usual notations of empirical risk (i.e. $\mathrm{R}_{\hat{S}}$ and R_S) as for the examples of $S^{(i)}$ coming from T_u, we only have access to their pseudo-label which can be wrong as expressed in the proof. We also put the superscript of the sample $S^{(i)}$ as a superscript of the risk to avoid overcharged notations.

$$\Leftrightarrow \quad \gamma_T^{(j-1)} > 0.$$

Last line is obtained by the weak assumption $\gamma_S^{(i)} > 0$. □

THEOREM 8.2 ([HAB 13b]).– *Let $S^{(0)}$ be the original learning set made of N_S labeled source data and T_u a set of $N_T \geq N_S$ unlabeled target examples. Let $\mathcal{A}$ be an iterative domain adaptation algorithm that randomly changes, at each step i, $2k$ original source labeled points from $S^{(i)}$ by $2k$ pseudo-labeled target examples randomly drawn from T_u without replacement and infers at each step a weak domain adaptation hypothesis (according to definition 8.3). Let $h^{(\frac{N_S}{2k})}$ be the weak domain adaptation hypothesis learned by $\mathcal{A}$ after $\frac{N_S}{2k}$ such iterations needed to change $S^{(0)}$ into a new learning set made of only target examples. Algorithm $\mathcal{A}$ solves a domain adaptation task with $h^{(\frac{N_S}{2k})}$ if*

$$\gamma_S^{(i)} \quad \geq \quad \gamma_T^{(i)}, \quad \forall i = 1, \ldots, \frac{N_S}{2k}, \qquad [8.1]$$

$$\gamma_S^{max} \quad > \quad \sqrt{\frac{\gamma_T^{(0)}}{2}}, \text{ where } \gamma_S^{max} = \max\left(\gamma_S^{(0)}, \ldots, \gamma_S^{(n)}\right). \qquad [8.2]$$

PROOF.– The hypothesis $h^{(\frac{N_S}{2k})}$ performs an actual domain adaptation if and only if $\mathrm{R}_{\hat{\mathcal{S}}}^{(\frac{N_S}{2k})}(h^{(\frac{N_S}{2k})}) < \hat{\mathrm{R}}_{\mathcal{T}}^{(0)}(h^{(0)})$. Roughly speaking, this condition simply means that the hypothesis $h^{(\frac{N_S}{2k})}$ learned only from pseudo-labeled target examples must outperform $h^{(0)}$ which have been learned only from source labeled data. Let $\hat{\mathrm{R}}_{\mathcal{S}_{2k}}^{(j)}(h^{(i)})$ be the error made by $h^{(i)}$ over the $2k$ pseudo-labeled examples inserted in the learning set at step j.

$$\begin{aligned}\hat{\mathrm{R}}_{\mathcal{S}_{2k}}^{(j)}(h^{(i)}) &= \left(\frac{1}{2} + \gamma_S^{(i)}\right)\left(\frac{1}{2} - \gamma_T^{(j-1)}\right) + \left(\frac{1}{2} - \gamma_S^{(i)}\right)\left(\frac{1}{2} + \gamma_T^{(j-1)}\right) \\ &= \frac{1}{2} - 2\gamma_S^{(i)}\gamma_T^{(j-1)}.\end{aligned}$$

We deduce that:

$$\mathrm{R}_{\hat{\mathcal{S}}}^{(\frac{N_S}{2k})}(h^{(\frac{N_S}{2k})}) < \hat{\mathrm{R}}_{\mathcal{T}}^{(0)}(h^{(0)})$$

$$\Leftrightarrow \frac{1}{N_S}\sum_{j=1}^{\frac{N_S}{2k}} 2k\left(\frac{1}{2} - 2\gamma_S^{(\frac{N_S}{2k})}\gamma_T^{(j-1)}\right) < \frac{1}{2} - \gamma_T^{(0)}$$

$$\Leftrightarrow \frac{4k}{N_S} \gamma_S^{(\frac{N_S}{2k})} \sum_{j=1}^{\frac{N_S}{2k}} \gamma_T^{(j-1)} > \gamma_T^{(0)}$$

$$\Leftrightarrow \frac{4k}{N_S} \gamma_S^{(\frac{N_S}{2k})} \sum_{j=1}^{\frac{N_S}{2k}} \gamma_S^{(j-1)} > \gamma_T^{(0)} \qquad [A5.1]$$

$$\Leftrightarrow \frac{4k}{N_S} \frac{N_S}{2k} (\gamma_S^{\max})^2 > \gamma_T^{(0)} \qquad [A5.2]$$

$$\Leftrightarrow \gamma_S^{\max} > \sqrt{\frac{\gamma_T^{(0)}}{2}},$$

where $\gamma_S^{\max} = \max(\gamma_S^{(0)}, \ldots, \gamma_S^{(n)})$. □

THEOREM 8.3 ([HAB 13b]).– *If condition [8.1] of theorem 8.2 is verified, then for all i,*

$$\tilde{\mathrm{R}}_S^{(i+1)}(h^{(i)}) > \tilde{\mathrm{R}}_S^{(i)}(h^{(i)}).$$

PROOF.– We have

$$\begin{aligned}
& \gamma_S^{(i)} \geq \gamma_T^{(i)} \\
\Leftrightarrow \quad & \frac{2k}{N_S}\gamma_S^{(i)} \geq \frac{2k}{N_S}\gamma_T^{(i)} \\
\Leftrightarrow \quad & \left(\frac{1}{2} - \gamma_S^{(i)}\right) + \frac{2k}{N_S}\gamma_S^{(i)} \geq \left(\frac{1}{2} - \gamma_S^{(i)}\right) + \frac{2k}{N_S}\gamma_T^{(i)} \\
\Leftrightarrow \quad & \left(\frac{1}{2} - \gamma_S^{(i)}\right) + \frac{2k}{N_S}\gamma_S^{(i)} - \frac{2k}{N_S}\gamma_T^{(i)} \geq \frac{1}{2} - \gamma_S^{(i)} \\
\Leftrightarrow \quad & \left(1 - \frac{2k}{N_S}\right)\left(\frac{1}{2} - \gamma_S^{(i)}\right) + \frac{2k}{N_S}\left(\frac{1}{2} - \gamma_T^{(i)}\right) \geq \frac{1}{2} - \gamma_S^{(i)} \\
\Leftrightarrow \quad & \tilde{\mathrm{R}}_S^{(i+1)}(h^{(i)}) \geq \tilde{\mathrm{R}}_S^{(i)}(h^{(i)}),
\end{aligned}$$

where $\tilde{\mathrm{R}}_S^{(i+1)}(h^{(i)})$ is the empirical error according to the (possibly wrong) labels at the disposal of the learner for the classifier $h^{(i)}$ over the new training set $S^{(i+1)}$ obtained once $2k$ new target pseudo-labeled data have been inserted. □

References

[ALQ 16] ALQUIER P., RIDGWAY J., CHOPIN N., "On the properties of variational approximations of gibbs posteriors", *Journal of Machine Learning Research*, JMLR.org, vol. 17, no. 1, pp. 8374–8414, 2016.

[AMB 06] AMBROLADZE A., PARRADO-HERNÁNDEZ E., SHAWE-TAYLOR J., "Tighter PAC-Bayes bounds", *Proceedings of the Conference on Neural Information Processing Systems (NIPS)*, pp. 9–16, 2006.

[BAR 02] BARTLETT P.L., MENDELSON S., "Rademacher and Gaussian complexities: risk bounds and structural results", *Journal of Machine Learning Research*, vol. 3, pp. 463–482, 2002.

[BAT 13] BATU T., FORTNOW L., RUBINFELD R. *et al.*, "Testing closeness of discrete distributions", *Journal of the ACM*, vol. 60, no. 1, pp. 1–25, 2013.

[BEC 13] BECKER C.J., CHRISTOUDIAS C.M., FUA P., "Non-linear domain adaptation with boosting", *Proceedings of the Conference on Neural Information Processing Systems (NIPS)*, pp. 485–493, 2013.

[BEL 13] BELLET A., HABRARD A., SEBBAN M., "A survey on metric learning for feature vectors and structured data", *arXiv preprint arXiv:1306.6709*, 2013.

[BEL 15] BELLET A., HABRARD A., SEBBAN M., *Metric Learning*, Morgan & Claypool Publishers, San Rafael, California, USA, 2015.

[BEN 07] BEN-DAVID S., BLITZER J., CRAMMER K. *et al.*, "Analysis of representations for domain adaptation", *Proceedings of the Conference on Neural Information Processing Systems (NIPS)*, pp. 137–144, 2007.

[BEN 09] BENGIO Y., "Learning deep architectures for AI", *Foundations and Trends® in Machine Learning*, Now Publishers, Inc., vol. 2, no. 1, pp. 1–127, 2009.

[BEN 10a] BEN-DAVID S., BLITZER J., CRAMMER K. *et al.*, "A theory of learning from different domains", *Machine Learning*, Springer, vol. 79, nos 1–2, pp. 151–175, 2010.

[BEN 10b] BEN-DAVID S., LU T., LUU T. *et al.*, "Impossibility theorems for domain adaptation", *Proceedings of the International Conference on Artificial Intelligence and Statistics (AISTATS)*, vol. 9, pp. 129–136, 2010.

[BEN 12a] BEN-DAVID S., LOKER D., SREBRO N. *et al.*, "Minimizing the misclassification error rate using a surrogate convex loss", *Proceedings of the International Conference on Machine Learning (ICML)*, pp. 83–90, 2012.

[BEN 12b] BEN-DAVID S., SHALEV-SHWARTZ S., URNER R., "Domain adaptation – can quantity compensate for quality?", *International Symposium on Artificial Intelligence and Mathematics (ISAIM)*, Fort Lauderdate, Florida, USA, 2012.

[BEN 12c] BEN-DAVID S., URNER R., "On the hardness of domain adaptation and the utility of unlabeled target samples", *Proceedings of the Conference on Algorithmic Learning Theory (ALT)*, pp. 139–153, 2012.

[BEN 13] BEN-DAVID S., URNER R., "Domain adaptation as learning with auxiliary information", *Workshop @ NIPS New Directions in Transfer and Multi-Task*, 2013.

[BIS 06] BISHOP C.M., *Pattern Recognition and Machine Learning*, Springer, Berlin, Heidelberg, Germany, 2006.

[BLA 10] BLANCHARD G., LEE G., SCOTT C., "Semi-supervised novelty detection", *Journal of Machine Learning Research*, vol. 11, pp. 2973–3009, 2010.

[BLI 08] BLITZER J., CRAMMER K., KULESZA A. *et al.*, "Learning bounds for domain adaptation", *Proceedings of the Conference on Neural Information Processing Systems (NIPS)*, pp. 129–136, 2008.

[BLU 98] BLUM A., MITCHELL T., "Combining labeled and unlabeled data with co-training", *Proceedings of the Conference on Learning Theory (COLT)*, pp. 92–100, 1998.

[BOH 14] BOHNÉ J., YING Y., GENTRIC S. *et al.*, "Large margin local metric learning", *Proceedings of the European Conference on Computer Vision (ECCV)*, pp. 679–694, 2014.

[BOL 07] BOLLEY F., GUILLIN A., VILLANI C., "Quantitative concentration inequalities for empirical measures on non-compact spaces", *Probability Theory and Related Fields*, Springer, vol. 137, nos 3–4, pp. 541–593, 2007.

[BOS 92] BOSER B.E., GUYON I.M., VAPNIK V.N., "A training algorithm for optimal margin classifiers", *Proceedings of the Annual Workshop on Computational Learning Theory (COLT)*, pp. 144–152, 1992.

[BOU 02] BOUSQUET O., ELISSEEFF A., "Stability and generalization", *Journal of Machine Learning Research*, vol. 2, pp. 499–526, 2002.

[BRU 10] BRUZZONE L., MARCONCINI M., "Domain adaptation problems: a DASVM classification technique and a circular validation strategy", *IEEE Transactions on Pattern Analysis and Machine Intelligence*, IEEE, vol. 32, no. 5, pp. 770–787, 2010.

[CAO 13] CAO Q., YING Y., LI P., "Similarity metric learning for face recognition", *Proceedings of the IEEE International Conference on Computer Vision (ICCV)*, pp. 2408–2415, 2013.

[CAT 07] CATONI O., *Lecture Notes – Monograph Series*, "PAC-Bayesian supervised classification: the thermodynamics of statistical learning", vol. 56, Institute of Mathematical Statistic, Beachwood, Ohio, USA, 2007.

[CHE 09] CHEN B., LAM W., TSANG I. *et al.*, "Extracting discriminative concepts for domain adaptation in text mining", *Proceedings of the ACM SIGKDD International Conference on Knowledge Discovery and Data Mining*, pp. 179–188, 2009.

[COR 11] CORTES C., MOHRI M., "Domain adaptation in regression", *Proceedings of the Conference on Algorithmic Learning Theory (ALT)*, pp. 308–323, 2011.

[COR 14] CORTES C., MOHRI M., "Domain adaptation and sample bias correction theory and algorithm for regression", *Theoretical Computer Science*, Elsevier, vol. 519, pp. 103–126, 2014.

[COR 15] CORTES C., MOHRI M., MUÑOZ MEDINA A., "Adaptation algorithm and theory based on generalized discrepancy", *Proceedings of the ACM SIGKDD International Conference on Knowledge Discovery and Data Mining*, pp. 169–178, 2015.

[COU 17] COURTY N., FLAMARY R., TUIA D., "Alain Rakotomamonjy: optimal transport for domain adaptation", *IEEE Transactions on Pattern Analysis and Machine Intelligence*, vol. 39, no. 9, pp. 1853–1865, 2017.

[COV 67] COVER T., HART P., "Nearest neighbor pattern classification", *IEEE Transactions on Information Theory*, Menlo Park, California, USA, vol. 13, no. 1, pp. 21–27, 1967.

[DAI 07] DAI W., YANG Q., XUE G.-R. *et al.*, "Boosting for transfer learning", *Proceedings of the International Conference on Machine Learning (ICML)*, pp. 193–200, 2007.

[DAV 07] DAVIS J.V., KULIS B., JAIN P. *et al.*, "Information-theoretic metric learning", *Proceedings of the International Conference on Machine Learning (ICML)*, pp. 209–216, 2007.

[DIE 00] DIETTERICH T.G., "Ensemble methods in machine learning", *International Workshop on Multiple Classifier Systems*, Springer, Prague, Czech Republic, pp. 1–15, 2000.

[FEI 06] FEI-FEI L., FERGUS R., PERONA P., "One-shot learning of object categories", *IEEE Transactions on Pattern Analysis and Machine Intelligence*, vol. 28, no. 4, pp. 594–611, 2006.

[FOU 15] FOURNIER N., GUILLIN A., "On the rate of convergence in wasserstein distance of the empirical measure", *Probability Theory and Related Fields*, Springer, vol. 162, nos 3–4, pp. 707–738, 2015.

[FRE 96] FREUND Y., SCHAPIRE R.E., "Experiments with a new boosting algorithm", *Proceedings of the International Conference on Machine Learning (ICML)*, pp. 148–156, 1996.

[FRE 97] FREUD Y., SCHAPIRE R.E., "A decision-theoretic generalization of on-line learning and an application to boosting", *Journal of Computer and System Sciences*, Elsevier, vol. 55, no. 1, pp. 119–139, 1997.

[GAO 14] GAO Z., GALVAO A., "Minimum integrated distance estimation in simultaneous equation models", *arXiv preprint arXiv:1412.2143*, 2014.

[GEN 11] GENG B., TAO D., XU C., "DAML: domain adaptation metric learning", *IEEE Transactions on Image Processing*, vol. 20, no. 10, pp. 2980–2989, 2011.

[GER 09] GERMAIN P., LACASSE A., LAVIOLETTE F.O. *et al.*, "PAC-Bayesian learning of linear classifiers", *Proceedings of the International Conference on Machine Learning (ICML)*, pp. 353–360, 2009.

[GER 13] GERMAIN P., HABRARD A., LAVIOLETTE F.O. *et al.*, "A PAC-Bayesian approach for domain adaptation with specialization to linear classifiers", *Proceedings of the International Conference on Machine Learning (ICML)*, pp. 738–746, 2013.

[GER 15a] GERMAIN P., HABRARD A., LAVIOLETTE F.O. *et al.*, "PAC-Bayesian theorems for domain adaptation with specialization to linear classifiers", *arXiv preprint arXiv:1503.06944*, 2015.

[GER 15b] GERMAIN P., LACASSE A., LAVIOLETTE F.O. *et al.*, "Risk bounds for the majority vote: from a PAC-Bayesian analysis to a learning algorithm", *Journal of Machine Learning Research*, JMLR.org, vol. 16, no. 1, pp. 787–860, 2015.

[GER 16] GERMAIN P., HABRARD A., LAVIOLETTE F.O. *et al.*, "A new PAC-Bayesian perspective on domain adaptation", *Proceedings of the International Conference on Machine Learning (ICML)*, vol. 48, pp. 859–868, 2016.

[GOP 11] GOPALAN R., LI R., CHELLAPPA R., "Domain adaptation for object recognition: an unsupervised approach", *Proceedings of the IEEE International Conference on Computer Vision (ICCV)*, pp. 999–1006, 2011.

[GRE 12] GRETTON A., BORGWARDT K.M., RASCH M.J. *et al.*, "A kernel two-sample test", *Journal of Machine Learning Research*, vol. 13, pp. 723–773, 2012.

[GRÜ 12] GRÜNEWÄLDER S., LEVER G., GRETTON A. *et al.*, "Conditional mean embeddings as regressors", *Proceedings of the International Conference on Machine Learning (ICML)*, Edinburgh, Scotland, pp.1823–1830, 2012.

[HAB 13a] HABRARD A., PEYRACHE J.-P., SEBBAN M., "Boosting for unsupervised domain adaptation", *Proceedings of the European Conference on Machine Learning and Principles and Practice of Knowledge Discovery in Databases (ECML/PKDD)*, pp. 433–448, 2013.

[HAB 13b] HABRARD A., PEYRACHE J.-P., SEBBAN M., "Iterative self-labeling domain adaptation for linear structured image classification", *International Journal on Artificial Intelligence Tools (IJAIT)*, World Scientific, vol. 22, no. 5, 2013.

[HAB 16] HABRARD A., PEYRACHE J.-P., SEBBAN M., "A new boosting algorithm for provably accurate unsupervised domain adaptation", *Knowledge and Information Systems*, Springer, vol. 47, no. 1, pp. 45–73, 2016.

[HAR 12] HAREL M., MANNOR S., "The perturbed variation", *Proceedings of the Conference on Neural Information Processing Systems (NIPS)*, pp. 1943–1951, 2012.

[HOE 63] HOEFFDING W., "Probability inequalities for sums of bounded random variables", *Journal of the American Statistical Association*, vol. 58, no. 301, pp. 13–30, 1963.

[HUA 06] HUANG J., SMOLA A.J., GRETTON A. *et al.*, "Correcting sample selection bias by unlabeled data", *Proceedings of the Conference on Neural Information Processing Systems (NIPS)*, pp. 601–608, 2006.

[JIN 09] JIN R., WANG S., ZHOU Y., "Regularized distance metric learning: theory and algorithm", *Proceedings of the Conference on Neural Information Processing Systems (NIPS)*, pp. 862–870, 2009.

[KIF 04] KIFER D., BEN-DAVID S., GEHRKE J., "Detecting change in data streams", *Proceedings of the International Conference on Very Large Data Bases*, pp. 180–191, 2004.

[KOL 59] KOLMOGOROV A.N., TIKHOMIROV V.M., "ε-entropy and ε-capacity of sets in function spaces", *Uspekhi Matematicheskikh Nauk*, Russian Academy of Sciences, Branch of Mathematical Sciences, vol. 14, no. 2, pp. 3–86, 1959.

[KOL 99] KOLTCHINSKII V., PANCHENKO D., "Rademacher processes and bounding the risk of function learning", *High Dimensional Probability II*, vol. 47, pp. 443–459, 1999.

[KUL 13] KULIS B., "Metric learning: a survey", *Foundations and Trends in Machine Learning*, vol. 5, no. 4, pp. 287–364, 2013.

[KUZ 13] KUZBORSKIJ I., ORABONA F., "Stability and hypothesis transfer learning", *Proceedings of the International Conference on Machine Learning (ICML)*, pp. 942–950, 2013.

[KUZ 17] KUZBORSKIJ I., ORABONA F., "Fast rates by transferring from auxiliary hypotheses", *Machine Learning*, Springer, vol. 106, no. 2, pp. 171–195, 2017.

[KUZ 18] KUZBORSKIJ I., Theory and algorithms for hypothesis transfer learning, PhD Thesis, EPFL, Lausanne, Switzerland, https://infoscience.epfl.ch/record/232494/files/EPFL_TH8011.pdf, 2018.

[LAC 06] LACASSE A., LAVIOLETTE F.O., MARCHAND M. *et al.*, "PAC-Bayes bounds for the risk of the majority vote and the variance of the gibbs classifier", *Proceedings of the Conference on Neural Information Processing Systems (NIPS)*, pp. 769–776, 2006.

[LAN 02] LANGFORD J., SHAWE-TAYLOR J., "PAC-Bayes & margins", *Proceedings of the Conference on Neural Information Processing Systems (NIPS)*, pp. 439–446, 2002.

[LAN 05] LANGFORD J., "Tutorial on practical prediction theory for classification", *Journal of Machine Learning Research*, vol. 6, pp. 273–306, 2005.

[LED 91] LEDOUX M., TALAGRAND M., *Probability in Banach Spaces: Isoperimetry and Processes*, Springer, Berlin Heidelberg, Germany, 1991.

[LEE 13] LEE D.-H., "Pseudo-label: the simple and efficient semi-supervised learning method for deep neural networks", *WorkshopICML on Challenges in Representation Learning*, Atlanta, USA, vol. 3, 2013.

[LET 59] LETTVIN J.Y., MATURANA H.R., MCCULLOCH W.S. *et al.*, "What the frog's eye tells the frog's brain", *Proceedings of the IRE*, vol. 47, no. 11, pp. 1940–1951, 1959.

[LI 07] LI X., BILMES J., "A bayesian divergence prior for classiffier adaptation", *Proceedings of the International Conference on Artificial Intelligence and Statistics (AISTATS)*, pp. 275–282, 2007.

[MAN 08] MANSOUR Y., MOHRI M., ROSTAMIZADEH A., "Domain adaptation with multiple sources", *Proceedings of the Conference on Neural Information Processing Systems (NIPS)*, pp. 1041–1048, 2008.

[MAN 09a] MANSOUR Y., MOHRI M., ROSTAMIZADEH A., "Domain adaptation: learning bounds and algorithms", *Proceedings of the Conference on Learning Theory (COLT)*, Montreal, Canada, 2009.

[MAN 09b] MANSOUR Y., MOHRI M., ROSTAMIZADEH A., "Multiple source adaptation and the Rényi divergence", *Proceedings of the Conference on Uncertainty in Artificial Intelligence (UAI)*, pp. 367–374, 2009.

[MAN 14] MANSOUR Y., SCHAIN M., "Robust domain adaptation", *Annals of Mathematics and Artificial Intelligence*, Springer, vol. 71, no. 4, pp. 365–380, 2014.

[MCA 99] MCALLESTER D.A., "Some PAC-Bayesian theorems", *Machine Learning*, Springer Netherlands, vol. 37, pp. 355–363, 1999.

[MCA 11] MCALLESTER D.A., KESHET J., "Generalization bounds and consistency for latent structural probit and ramp loss", *Proceedings of the Conference on Neural Information Processing Systems (NIPS)*, pp. 2205–2212, 2011.

[MCN 17] MCNAMARA D., BALCAN M., "Risk bounds for transferring representations with and without fine-tuning", *Proceedings of the International Conference on Machine Learning (ICML)*, pp. 2373–2381, 2017.

[MIT 97] MITCHELL T.M., *Machine learning*, McGraw-Hill, Inc., New York, USA, 1997.

[MOH 12] MOHRI M., ROSTAMIZADEH A., TALWALKAR A., *Foundations of Machine Learning*, MIT Press, Cambridge, Massachusetts, USA, 2012.

[MON 81] MONGE G., "Mémoire sur la théorie des déblais et des remblais", *Histoire de l'Académie Royale des Sciences*, pp. 666–704, 1781.

[MOR 12a] MORENO-TORRES J.G., RAEDER T., ALAIZ-RODRÍGUEZ R. *et al.*, "A unifying view on dataset shift in classification", *Pattern Recognition*, Elsevier, vol. 45, no. 1, pp. 521–530, 2012.

[MOR 12b] MORVANT E., HABRARD A., AYACHE S., "Parsimonious unsupervised and semi-supervised domain adaptation with good similarity functions", *Knowledge and Information Systems*, Springer, vol. 33, no. 2, pp. 309–349, 2012.

[MOR 15] MORVANT E., "Domain adaptation of weighted majority votes via perturbed variation-based self-labeling", *Pattern Recognition Letters*, Elsevier, vol. 51, pp. 37–43, 2015.

[MÜL 97] MÜLLER A., "Integral probability metrics and their generating classes of functions", *Advances in Applied Probability*, Cambridge University Press, UK, vol. 29, no. 2, pp. 429–443, 1997.

[ORA 09] ORABONA F., CASTELLINI C., CAPUTO B. *et al.*, "Model adaptation with least-squares SVM for adaptive hand prosthetics", *Proceedings of the IEEE International Conference on Robotics and Automation*, pp. 2897–2903, 2009.

[PAN 08] PAN S.J., KWOK J.T., YANG Q., "Transfer learning via dimensionality reduction", *Proceedings of the National Conference on Artificial Intelligence (AAAI)*, pp. 677–682, 2008.

[PAN 09] PAN S.J., TSANG I.W., KWOK J.T. *et al.*, "Domain adaptation via transfer component analysis", *Proceedings of the International Joint Conference on Artificial Intelligence (IJCAI)*, pp. 1187–1192, 2009.

[PAR 10] PARAMESWARAN S., WEINBERGER K.Q., "Large margin multi-task metric learning", *Proceedings of the Conference on Neural Information Processing Systems (NIPS)*, pp. 1867–1875, 2010.

[PAR 12] PARRADO-HERNÁNDEZ E., AMBROLADZE A., SHAWE-TAYLOR J. *et al.*, "PAC-Bayes bounds with data dependent priors", *Journal of Machine Learning Research*, vol. 13, pp. 3507–3531, 2012.

[PÉR 07] PÉREZ Ó., SÁNCHEZ-MONTAÑÉS M., "A new learning strategy for classification problems with different training and test distributions", *International Work-Conference on Artificial Neural Networks*, Springer, San Sebastián, Spain, pp. 178–185, 2007.

[PER 15a] PERROT M., HABRARD A., "Regressive virtual metric learning", *Proceedings of the Conference on Neural Information Processing Systems (NIPS)*, pp. 1810–1818, 2015.

[PER 15b] PERROT M., HABRARD A., "A theoretical analysis of metric hypothesis transfer learning", *Proceedings of the International Conference on Machine Learning (ICML)*, pp. 1708–1717, 2015.

[PER 16] PERROT M., COURTY N., FLAMARY R. *et al.*, "Mapping estimation for discrete optimal transport", *Proceedings of the Conference on Neural Information Processing Systems (NIPS)*, pp. 4197–4205, 2016.

[RE 12] RE M., VALENTINI G., "Ensemble methods: a review", *Advances in Machine Learning and Data Mining for Astronomy*, Chapter 26, pp. 563–582, 2012.

[RED 15] REDKO I., Nonnegative matrix factorization for unsupervised transfer learning, PhD Thesis, Paris North University, France, 2015.

[RED 17] REDKO I., HABRARD A., SEBBAN M., "Theoretical analysis of domain adaptation with optimal transport", *Proceedings of the European Conference on Machine Learning and Principles and Practice of Knowledge Discovery in Databases (ECML/PKDD)*, pp. 737–753, 2017.

[ROS 58] ROSENBLATT F., "The Perceptron", *Psychological Review*, vol. 65, no. 6, pp. 386–408, 1958.

[ROY 16] ROY J.-F., MARCHAND M., LAVIOLETTE F.O., "A column generation bound minimization approach with PAC-Bayesian generalization guarantees", *Proceedings of the International Conference on Artificial Intelligence and Statistics (AISTATS)*, pp. 1241–1249, 2016.

[SAE 10] SAENKO K., KULIS B., FRITZ M. *et al.*, "Adapting visual category models to new domains", *Proceedings of the European Conference on Computer Vision (ECCV)*, pp. 213–226, 2010.

[SAI 97] SAITOH S., *Integral Transforms, Reproducing Kernels and their Applications*, Pitman Research Notes in Mathematics Series, Longman, Harlow, UK, 1997.

[SAI 17] SAITO K., USHIKU Y., HARADA T., "Asymmetric tri-training for unsupervised domain adaptation", *Proceedings of the International Conference on Machine Learning (ICML)*, pp. 2988–2997, 2017.

[SAN 14] SANDERSON T., SCOTT C., "Class proportion estimation with application to multiclass anomaly rejection", *Proceedings of the International Conference on Artificial Intelligence and Statistics (AISTATS)*, pp. 850–858, 2014.

[SAU 98] SAUNDERS C., GAMMERMAN A., VOVK V., "Ridge regression learning algorithm in dual variables", *Proceedings of the International Conference on Machine Learning (ICML)*, pp. 515–521, 1998.

[SCH 99] SCHAPIRE R.E., "A brief introduction to boosting", *Proceedings of the International Joint Conference on Artificial Intelligence (IJCAI)*, pp. 1401–1406, 1999.

[SCO 13] SCOTT C., BLANCHARD G., HANDY G., "Classification with asymmetric label noise: consistency and maximal denoising", *Proceedings of the Conference on Learning Theory (COLT)*, pp. 489–511, 2013.

[SEE 02] SEEGER M., "PAC-Bayesian generalisation error bounds for gaussian process classification", *Journal of Machine Learning Research*, vol. 3, pp. 233–269, 2002.

[SEJ 13] SEJDINOVIC D., SRIPERUMBUDUR B., GRETTON A. *et al.*, "Equivalence of distance-based and RKHS-based statistics in hypothesis testing", *The Annals of Statistics*, vol. 41, no. 5, pp. 2263–2291, 2013.

[SEN 16] SENER O., SONG H.O., SAXENA A. *et al.*, "Learning transferrable representations for unsupervised domain adaptation", *Proceedings of the Conference on Neural Information Processing Systems (NIPS)*, pp. 2110–2118, 2016.

[SHA 14] SHALEV-SHWARTZ S., BEN-DAVID S., *Understanding Machine Learning: From Theory to Algorithms*, Cambridge University Press, UK, 2014.

[SHI 00] SHIMODAIRA H., "Improving predictive inference under covariate shift by weighting the log-likelihood function", *Journal of Statistical Planning and Inference*, Elsevier, vol. 90, no. 2, pp. 227–244, 2000.

[SIN 67] SINKHORN R., KNOPP P., "Concerning nonnegative matrices and doubly stochastic matrices", *Pacific Journal of Mathematics*, Mathematical Sciences Publishers, vol. 21, no. 2, pp. 343–348, 1967.

[SON 08] SONG L., Learning via hilbert space embedding of distributions, PhD Thesis, University of Sydney, Australia, 2008.

[SUG 08] SUGIYAMA M., NAKAJIMA S., KASHIMA H. *et al.*, "Direct importance estimation with model selection and its application to covariate shift adaptation", *Proceedings of the Conference on Neural Information Processing Systems (NIPS)*, pp. 1433–1440, 2008.

[TOM 10] TOMMASI T., ORABONA F., CAPUTO B., "Safety in numbers: learning categories from few examples with multi model knowledge transfer", *Proceedings of the IEEE Conference on Computer Vision and Pattern Recognition (CVPR)*, pp. 3081–3088, 2010.

[VAL 84] VALIANT L.G., "A theory of the learnable", *Communications of the ACM*, vol. 27, pp. 1134–1142, 1984.

[VAP 71] VAPNIK V.N., CHERVONENKIS A.Y., "On the uniform convergence of relative frequencies of events to their probabilities", *Theory of Probability and its Applications*, SIAM, vol. 16, no. 2, pp. 264-280, 1971.

[VAP 95] VAPNIK V.N., *The Nature of Statistical Learning Theory*, Springer-Verlag, New York, USA, 1995.

[VAP 06] VAPNIK V., *Estimation of Dependences Based on Empirical Data*, Springer Science & Business Media, New York, USA, 2006.

[VIL 09] VILLANI C., *Optimal Transport: Old and New*, Grundlehren der mathematischen Wissenschaften, Springer, Berlin, Germany, 2009.

[WAN 14] WANG X., HUANG T.-K., SCHNEIDER J., "Active transfer learning under model shift", *Proceedings of the International Conference on Machine Learning (ICML)*, pp. II-1305–II-1313, 2014.

[XU 10] XU H., MANNOR S., "Robustness and generalization", *Proceedings of the Conference on Learning Theory (COLT)*, pp. 503–515, 2010.

[YAN 07] YANG J., YAN R., HAUPTMANN A.G., "Cross-domain video concept detection using adaptive SVMs", *Proceedings of the ACM International Conference on Multimedia (ACM Multimedia)*, pp. 188–197, 2007.

[ZHA 09] ZHA Z.J., MEI T., WANG M. *et al.*, "Robust distance metric learning with auxiliary knowledge", *Proceedings of the International Joint Conference on Artificial Intelligence (IJCAI)*, pp. 1327–1332, 2009.

[ZHA 12] ZHANG C., ZHANG L., YE J., "Generalization bounds for domain adaptation", *Proceedings of the Conference on Neural Information Processing Systems (NIPS)*, pp. 3320–3328, 2012.

[ZHA 13] ZHANG K., SCHÖLKOPF B., MUANDET K. *et al.*, "Domain adaptation under target and conditional shift", *Proceedings of the International Conference on Machine Learning (ICML)*, pp. 819–827, 2013.

[ZOL 84] ZOLOTAREV V.M., "Probability metrics", *Theory of Probability and Its Applications*, vol. 28, no. 2, pp. 278–302, 1984.

Index

Printed in the United States
By Bookmasters